D0426242

vestigations that have within a single decade revolutionized our understanding of life, but comes to understand, as well, the delicate and precise mechanisms by which sleep, reproduction, and physical function in every living organism respond to the rhythms of the solar day, the lunar month, the ocean tides, the terrestrial year.

Just as *Gods, Graves, and Scholars* made available to the general reader the excitement of archaeology, so the literally marvelous manifestations of biochronometry are now made clear and immediate in *The Living Clocks*.

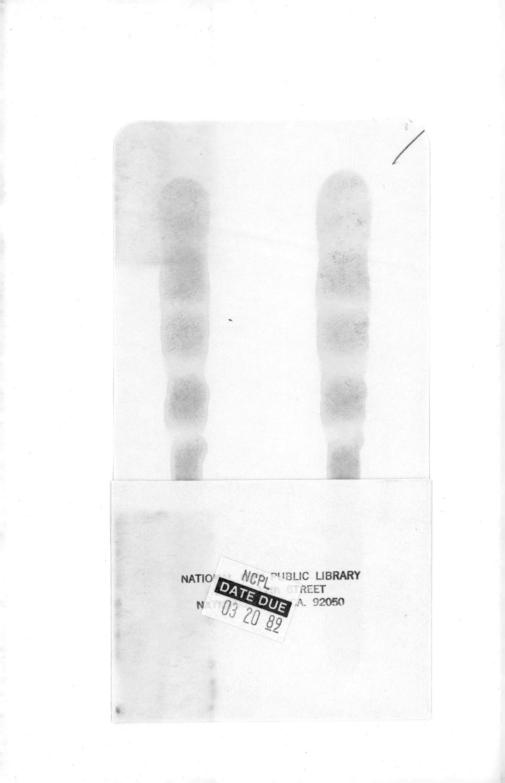

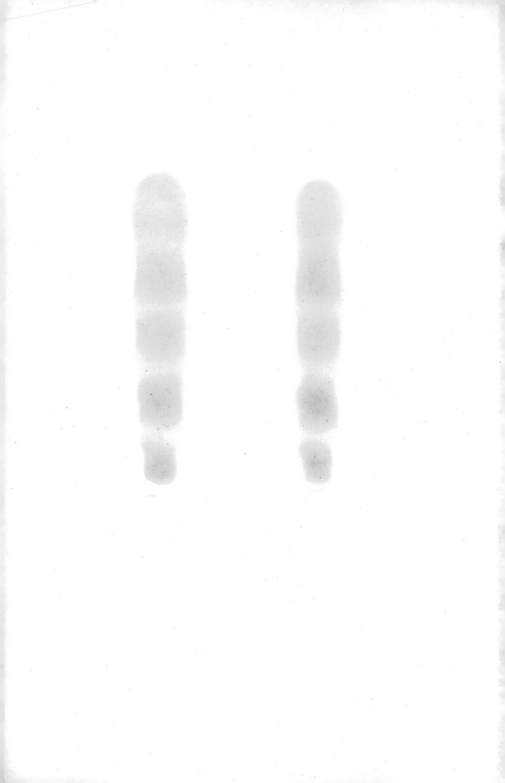

The Living Clocks

Drawings by Hollett Smith

THE LIVING CLOCKS

Ritchie R. Ward

Alfred A. Knopf · New York · 1971

TO CLAIRE

Preface

It is now more than seven years since I first became fascinated by living clocks and the scientists who study them. Yet in little more than that span, science has revolutionized its view of the rhythms that permit living things to tell time. What was once an interesting aspect of natural history has become a fundamental base—if not the central core—of modern biology.

No single idea of the twentieth century has contributed more broadly to today's understanding of the life sciences than that organisms—from one-celled plants to man—either are endowed with, or perhaps *are*, living clocks.

And certainly no idea of such far-reaching consequences has been less accessible to the general reader. The biologists who write of biochronometry, as they call it, have a specialized language of their own within the special language of biology itself, and it is hardly surprising that a cultural gap has separated the intelligent man from the momentous discoveries in this field. In this book I have tried to tell the story of living clocks for readers who have no special background in science, but with such attention to accuracy that no scientist will take offense. At the least, I hope that both will enjoy it.

The field is so broad that I have been much limited in my choice of subjects and personalities to write of. In a few cases, the choice was mandatory: when an authority had said that a discovery was "one of the dramatic zoological events of this century," then I wrote of that. Other choices were not that clear. My simple solution was to select those discoveries that seemed to me to be the most significant and the most interesting. Misjudgments in selection and emphasis are therefore entirely my own.

A book of this kind would be out of the question without the help of many others. My own debt is great to those who had the vision and energy to make possible the Cold Spring Harbor Symposium on Biological Clocks, held at the Biological Laboratory of the Long Island Biological Association, in the spring of 1960. Acknowledgment of these contributors has been made in the Foreword to the Symposium volume, published by the Association in 1961. There, Arthur Chovnik expressed the hope that the Symposium would "serve as a unifying influence on the entire field of study." It did.

More directly, I am in debt to Michael Menaker of the University of Texas, Chairman of the Organizing Committee for the International Biochronometry Symposium held at Friday Harbor, Washington, in the fall of 1969. In the face of an overflow of advance reservations, he invited me to come too.

Many individual scientists have been generous in bearing with long exchanges of correspondence, giving the time to answer endless questions in interviews, and explaining and demonstrating their research methods and equipment. I cannot name them all, but special thanks go to Jürgen Aschoff, Max-Planck-Institute, Erling-Andechs, West Germany; Harry A. Borthwick and Sterling B. Hendricks of the U.S. Department of Agriculture, Beltsville, Maryland; Frank A. Brown, Jr., Northwestern University; Erwin Bünning, University of Tübingen, West Germany; J. L. Cloudsley-Thompson, University of Khartoum; Patricia DeCoursey, University of South Carolina; Charles F. Ehret, Argonne National Laboratory; Lloyd T. Evans, Commonwealth Scientific and Industrial Research Organization, Canberra; William J. Hamilton, University of California, Davis; Karl C. Hamner, UCLA; Janet Harker and E. A. C. MacRobbie, Girton College, Cambridge; Hudson Hoagland, Worcester (Massachusetts) Foundation for Experimental Biology; Robert G. Lindberg, Northrop Corporation; Colin S. Pittendrigh, Stanford University; Frank B. Salisbury, Utah State University; E. G. Franz Sauer, Zoological Institute, Bonn, West Germany; Hans Selye, University of Montreal; Ursula v. St. Paul, Max-Planck-Institute, Erling-Andechs, West Germany; Beatrice M. Sweeney, University of California, Santa Barbara; and to Lowell Thomas, who told me how it feels to have one's own biological clock disrupted. Special thanks too, to Hollett Smith, for the imagination of her drawings that enliven the text. Much as they have helped, none of these people are in any way responsible for such errors as I may have made in fact or interpretation.

My editor, Harold Strauss, has borne with me for long. Without his patience and encouragement, the book would have been impossible.

R. R. W.

Orinda, California

Contents

Illustrations

The Living Clocks

There seems to be little doubt
that future thinking in
all the biological sciences
will have to be done in
terms of periodic variations
if we are to make good sense.

WILLIAM WOLF,

1962

1. The Rhythms of Life

From the dawn of modern natural philosophy, which rose with Galileo's overthrow of Aristotelian physics, science has advanced in great forward bursts rather than at a uniform rate. The great surges have been stimulated either by revolutionary new concepts or by new techniques. James B. Conant has vividly put it: "Tremendous spurts in the progress of the various sciences are almost always connected with the development of a new technique or the sudden emergence of a new concept. It is as though a group of prospectors were hunting on barren ground and suddenly struck a rich vein of ore. All at once everyone begins to work feverishly and the gold begins to flow."[1]

[1] James B. Conant: *On Understanding Science* (New Haven: Yale University Press; 1947), pp. 73–4.

3

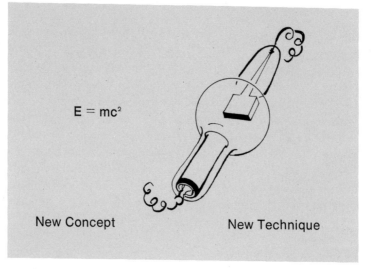

$E = mc^2$

New Concept   New Technique

Revolutionary new concepts and techniques paved the way toward great forward surges in science. Such a new concept was Einstein's famous mass-energy equation, in which E is energy, m is mass, and c² is the square of the speed of light, a number so large as to be virtually incomprehensible. Such a new technique was the cathode ray tube that Roentgen used in his accidental discovery of X rays.

In the first half of the twentieth century the great forward surges were made in physics. A classic example of a new concept that led to feverish activity was Einstein's famous equation showing that when mass is converted to energy, the energy released is proportional to the square of the speed of light. This concept stimulated not only a flood of new experimental work in basic physics but, as the century progressed, also led to a new technology destined to revolutionize the arts of war and peace. Parallel to this new concept, a far-reaching new technique in physics was the discovery of X rays. "In our own day after the publication of an account of the X-ray tube, new experimental facts came forth in torrents."[2]

Today it seems virtually certain that the great forward surges of the second half of the twentieth century will be made

[2] *Ibid.*, p. 73.

in biology. In 1953 James D. Watson and Francis H. C. Crick formulated a theory of the structure of the DNA molecule that accounted for the fact that the gene—the determiner of heredity in living organisms—can replicate itself exactly. This new concept opened floodgates of research in biology and furnished the key to deeper understandings of the processes of life itself. Now physiologists, biochemists, biophysicists, and bioengineers are studying how breakdowns in the basic life processes can lead to hereditary defects, as well as to aging, cancer, and mental illness. To understand the mechanism of these processes more clearly is to come closer to the promise of cure.[3]

At about the same time that Watson and Crick were working out the structure of the DNA molecule at Cambridge University in England, Gustav Kramer was studying the navigation of birds at the Max-Planck-Institute in Germany. In a brilliant series of experiments he was able to show that birds achieve their incredibly accurate directional flight through their ability to navigate by the sun. Even more significant, he showed that birds correct for the sun's continuous movement by referring to an internal mechanism that can be likened to a clock. The analogy to a ship's navigator using the modern man-made sextant and chronometer is very close.

Since Kramer's work, the term "biological clock" has come into general use among scientists, and a torrent of experimental work yielding unexpected new insights into age-old biological problems has flowed from this concept. Throughout centuries, innumerable observations have been made indicating that plants and animals carry out their activities in regularly timed rhythms, but these were no more than interesting facts of nature until the emergence of the concept of the clock. This concept was an intellectual springboard that stimulated fruitful experimental programs in many fields.

From the earliest times it has been known that plants and animals display periodically recurring activities. The daily

[3] The general reader will find an excellent account of the great discoveries in this field in Ruth Moore: *The Coil of Life* (New York: Alfred A. Knopf, Inc.; 1962).

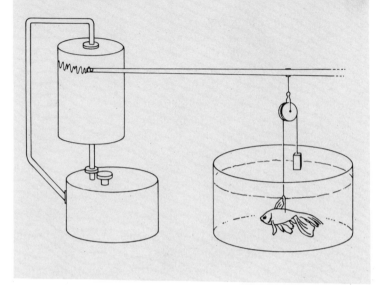

The common goldfish exhibits a marked daily rhythm that persists in constant laboratory conditions. J. S. Szymanski showed this as early as 1914. The aktograph he used recorded the periodic activity of the goldfish by means of a pointer writing on a revolving smoked drum.

sleeping and waking cycle of men, animals, and some plants was obvious, as were the monthly calendar cycles—especially those timed to the lunar rather than to the calendar month. With the advent of the clock concept, new observations were motivated, older observations fell into more logical relationships, and it was discovered that the recurring activities of living things have periods differing over a very wide range. Some have been found to occur as often as every thousandth of a second; others occur every second or every hour. The most widely studied rhythms have periods of 24 hours, but this is only approximate, and may be as short as 22 hours or as long as 26. Then there are weekly rhythms and monthly rhythms, and in some organisms the conspicuous periodic event may occur only every seven years, or even every seventeen.

Whether an organism finds it biologically useful to repeat an event every thousandth of a second or every seventeen years, it is clear that it could function efficiently only if it were provided with some means of measuring the passage of time. If, therefore, plants and animals do have within themselves an actual clock mechanism by which they are able to tell time, a flood of new questions arises. Where in the organism is the clock located? How does it work? Does it help the organism in its struggle for survival, and if so, how?

In searching for answers to these questions, biologists made an unexpected and revolutionary discovery. This was that an organism's response to an external stimulus varies critically with the phase of the rhythmic cycle that it happens to be in at the time. All previous observations of natural processes—all of the carefully planned laboratory experiments—had been made in ignorance of this fact.

As recently as 1962, William Wolf of New York University said, "We know that periodic variations may go so far that the same agent may have a favorable effect when given at one phase of a periodic cycle and may kill when given in another phase. We shall thus find that much of the current and past literature may well acquire entirely new meanings or even become meaningless when pursued in the light of biological rhythms. Many apparently inexplicable, subtle, or gross irregularities in the behavior of pharmaceutical, physical, chemical, enzymatic, and psychological agents are likely to become more consistent when functional rhythms are recognized. It may easily follow that much of the literature may have to be discarded or revised when normal and abnormal variations in rhythms are taken into account."[4]

The biology texts must be rewritten!

The new texts will note, for example, that nose and throat surgeons in Tallahassee, Florida, found that hemorrhages in

[4] William Wolf: "Introductory Remarks," *Rhythmic Functions in the Living System, Annals of the New York Academy of Sciences,* Vol. XCVIII (October 30, 1962), p. 755.

throat operations were 82 percent higher in the second quarter of the moon than at other times.[5] They will stress, in short, the critical fact that any biological experiment that fails to take the effects of natural cycles into account is not a controlled experiment at all,[6] and does not deserve to be called scientific in the modern sense.

Naturally the new texts will make no claim to providing the final answers. In physics and engineering new results are still evolving from Einstein's 1905 concept of the equivalence of mass and energy, and it is reasonable to suppose that new consequences of the biological clock concept will continue to evolve to the benefit of agriculture, medicine, and space engineering through A.D. 2000. Further, the texts of different authors will stress varying points of view. Biologists differ sharply today about how living clocks work, and the problem is so complex that a reconciliation seems hardly likely within the next few years.

According to J. L. Cloudsley-Thompson, an internationally recognized authority, biological rhythms—in particular the daily ones—might be explained in any one of three ways:

—They are learned.

—They are inherited.

—They depend on reactions to cosmic stimuli.[7]

Each of these hypotheses has its own ardent advocates, and there is not yet enough evidence for any of the proponents to convince the others that they are wrong. It may well be that each of these views is both right and wrong to some extent and that a general consensus will someday be reached through a reconciliation of all of them.

Today, however, the battle of ideas rages, and in later chapters we shall visit the laboratories and join the field ex-

---

[5] *Ibid.*, p. 756.

[6] Conant (*On Understanding Science*, pp. 50–7) has given a definitive statement of the function of the controlled experiment in the tactics and strategy of science. The importance of the controlled experiment in biology will be stressed throughout this book.

[7] J. L. Cloudsley-Thompson: *Rhythmic Activity in Animal Psysiology and Behaviour* (New York: Academic Press, Inc.; 1961), p. 199.

peditions of the leaders of the opposing forces. We shall live
the history of science in the making.

First, of course, we shall need to review briefly the historical background that has led up to the present conflict of
ideas, and how the opposing forces came to occupy their
present strategic positions. Let us first consider in turn each
of the three ways in which a living clock may be thought to
work, in the order given by Cloudsley-Thompson.

Speaking only of animals, W. H. Thorpe suggested before
1956 that a sense of time might be *learned* in a way similar to
the extraordinary process of imprinting, in which an animal
picks up an idea on the instant, and retains it virtually for life.[8]
The first animals known to behave in this way were young
geese—isolated except for the man who fed them. Within hours,
the idea of human beings as parents or friends was imprinted
on the minds of these impressionable young geese. After that
they spurned other geese and were happy only in the company
of men. As a consequence of this observation, Thorpe suggested
that biological rhythms might be set in motion by the very
early impressions young animals receive of the daily cycle of
the earth's rotation, and that once so started, the animal's
rhythmic processes would continue for life. For some reason,
biologists did not pursue this idea with much interest, and
little work has been published either to support or refute it.

The active controversy today revolves around the question
as to whether living clocks are *inherited* or depend on *cosmic
stimuli.* Cloudsley-Thompson is one of a number who believe
that biological clocks are inherited. "[This] theory, to which
the majority of workers in the field of rhythms now subscribes,
assumes the presence of an inherited 24-hour 'clock.' Considerable evidence is afforded for this view from the fact that certain
periodicities may be accelerated or retarded under 'constant'
laboratory conditions. It is argued that if subtle environmental
'residual periodic variables' were responsible for the rhythms
of the organism, they would surely have acted as 'clues' and

[8] W. H. Thorpe: *Learning and Instinct in Animals,* 2d ed. (Cambridge,
Mass.: Harvard University Press; 1963), pp. 128–31.

set up an *exact* 24-hour periodicity. But this is not conclusive . . ."[9]

A number of additional arguments have been advanced by those who believe that the clocks are inherited. One of these holds that the clocks could have reached their present high accuracy only through a process of natural selection. Those animals whose clocks ran either fast or slow would have dropped from the race. Another argument is that within rather wide limits, living clocks maintain their accuracy in spite of changes in temperature, like a chronometer that is compensated for temperature. What makes this hard to understand is that the clockwork must be driven by biochemical reactions, and the rates of these reactions double for every 10-degree rise in temperature. Yet, some temperature-compensating mechanism can be argued to fit neatly into the theory of natural selection.

Vigorously and articulately opposed to inheritance as the origin of the primary timer is Frank A. Brown, Jr., Morrison Professor of Biology at Northwestern University. Brown and his many co-workers, who have been studying animal and plant rhythms for more than three decades, are convinced that tides in forces from outer space are the master clocks that synchronize living clocks. These tides are bound to the phases of the moon and to sunspot cycles. They ebb and flow to many rhythms: hourly, daily, monthly, and yearly. In turn, they drive other tides in the many unseen geophysical forces that immediately surround the earth. These are forces like barometric pressure, the earth's magnetic field, ionization of the atmosphere, cosmic rays, and weak electromagnetic fields of radio frequency. All of these tides are so extremely weak that it seems incredible that any organism could sense slight changes in them.

A man, deep in a forest, tells north with a pocket compass. Yet if he were as sensitive to the earth's magnetic field as a mud snail, he would not need the compass. He could feel which way was north. Brown and his colleagues have shown beyond doubt that mud snails can sense magnetic fields as weak as those of the earth.

[9] Cloudsley-Thompson: *Rhythmic Activity*, pp. 9–10.

the important discoveries: their fumblings, their oversights, and their misinterpretations no less than their amazing laboratory skills and their triumphs of insight. At the same time we shall place in perspective the tactics and strategy of science, so that the nonspecialist may clearly appreciate some of the bizarre paths scientists have chosen in their efforts to unveil the secrets of nature.

But first we shall need to agree on what kinds of activities we shall be calling "science," and what kinds of activities we shall say are "not science." Concerning the latter, we shall say that the mere collection of data—however detailed or accurate —is not science. We shall also say that the mere classification of observed facts—however complete and logical—is not science. For what science is, we shall rely on James B. Conant's definition, even though he has offered it with the reservation that the logical analyses of the philosophers might yield a better one:

"As a first approximation, we say that science emerges from the other progressive activities of man to the extent that new concepts arise from experiments and observations, and the new concepts in turn lead to further experiments and observations. The case histories drawn from the last three hundred years show examples of fruitful and fruitless concepts. The texture of modern science is the result of the interweaving of the fruitful concepts. The test of a new idea is therefore not only its success in correlating the then-known facts but much more its success or failure in stimulating further experimentation or observation which in turn is fruitful. This dynamic quality of science viewed not as a practical undertaking but as a development of conceptual schemes seems to be close to the heart of the best definition. It is this quality which can be demonstrated only by the historical approach, or else learned by direct professional experience."[6]

With this understanding, we turn in the next chapter to

[6] Conant: *On Understanding Science*, p. 24.

In a brief summary of both sides of the controversy, Brown has written: "Two hypotheses have, historically, been entertained as to the nature of the period-timer for the mysterious, temperature-resistant period lengths of the clocks. One hypothesis is that through natural selection organisms have evolved an inherited complex of internal, biochemical oscillating systems, which parallel very closely all the natural geophysical periods of the organism's environment. The other hypothesis is that organisms, even when deprived of all the obvious environmental factors such as light and temperature, still oscillate with all the natural geophysical frequencies as a consequence of response to rhythmic changes in pervasive forces of the environment, and that these induced oscillations comprise the basic timers. Indeed, it has now been proven that living systems are astoundingly sensitive to very weak magnetostatic, electrostatic, and perhaps also weak electromagnetic fields of radio-frequency. The sensitivity is adequate for response to the earth's natural fields."[1]

Brown then gave his own views in support of the cosmic-stimulus hypothesis:

"While it may be proven in the future that organisms also have available fully independent timing systems for the natural geophysical periods, there is no evidence yet available to indicate it. All our current knowledge of the clocks of animals and plants falls into a consistent rational scheme in terms of biological 'clock works,' with remarkably similar properties throughout all plants and animals, beautifully adapted for maximal usefulness to the organisms, and dependent for their timing upon continuing response to the subtle geophysical environment."[2]

Janet Harker, whose brilliant work at Cambridge University first showed the actual location of the biological clock in any living organism, gave her evaluation of the current status of the inheritance vs. cosmic force theories in the fol-

[1] Frank A. Brown, Jr.: *Biological Clocks*, Biological Sciences Curriculum Study Pamphlet No. 2 (Boston: D. C. Heath and Company; 1962), p. 35.
[2] *Ibid.*, p. 35.

lowing words: ". . . Brown contends that at all times organisms are under the influence of environmental variables which could give time-cues. This contention is by no means accepted by the majority of those who work on rhythms, but so far no evidence whatever has been recorded which in any way disproves the effects observed by Brown."[3]

Answers to these basic questions are now being sought by scientists working in laboratories and in the field all over the world. Why should they pursue the quest with such intensity of purpose?

No doubt the most impelling reason arises from man's insatiable curiosity to learn more about the world and how it works. Another is the scientist's hope that at some point in his work he will discover something—very possibly unexpected and different from what he was looking for—that might prove to be of ultimate benefit to society.

Isaac Asimov has observed that man's curiosity drive can be ascribed to his need to keep his brain more fully occupied than he finds it necessary to provide himself with food, shelter, and other items of immediate usefulness: "Thus the desire to know seems to lead into successive realms of greater etherealization and more efficient occupation of the mind—from knowledge of accomplishing the useful, to knowledge of accomplishing the esthetic, to 'pure' knowledge.

"Knowledge for itself alone seeks answers to such questions as 'How high is the sky?' or 'Why does a stone fall?' This is sheer curiosity—curiosity at its idlest and therefore perhaps at its most peremptory. After all, it serves no apparent purpose to know how high the sky is or why the stone falls. The lofty sky does not interfere with the ordinary business of life, and as for the stone, knowing why it falls does not help us to dodge it more skillfully, or soften the blow if it happens to hit us. Yet there have always been people who ask such apparently use-

[3] Janet Harker: *The Physiology of Diurnal Rhythms* (Cambridge University Press; 1964), p. 88.

less questions and try to answer them out of the sheer desire to know—out of the absolute necessity of keeping the brain working."[4]

Once the "pure" scientist has discovered a new concept or a new set of facts arising from such a concept, the medical men, the agricultural experts, the engineers, and the other innovators in the practical arts take over and develop these ideas into tools, machines, and processes useful to society.

As the story of living clocks unfolds, we shall see how Janet Harker, seeking only to learn exactly where the clock is located in a cockroach, found that changing the regulation of the clock caused such severe stresses in the cockroaches that they promptly developed cancer and died; and we shall see how this discovery influenced the thinking of Franz Halberg, a cancer authority working at the University of Minnesota Medical School.

We shall see how the astonishing discovery of Frank A. Brown, Jr., that potatoes can predict the barometer several days in advance is now being used at the request of NASA as a technique to investigate how living organisms (including astronauts) may react to the environment of outer space. Through such comparisons, it will become apparent how advances in technology depend on advances in science, and of equal importance, how advances in science depend on advances in technology.[5] It is obvious that if space flight had not become a reality, no scientist would have had the opportunity to explore the fascinating question of man's chances of surviving his journey to the stars.

Throughout this book we shall be looking at the story of living clocks from a historical point of view—that is to s through the eyes and minds of the men and women who

[4] Isaac Asimov: *The Intelligent Man's Guide to Science* Basic Books, Inc.; 1960), p. 7.
[5] The late Professor J. L. Henderson used to tell his stu 1850 the steam engine did more for science than s steam engine.

a series of absorbing observations of the behavior of living clocks, which stopped short of being scientific, but which nevertheless led in the course of time to fruitful new ideas which in turn stimulated many rewarding explorations of the rhythms of life.

*There is a procedure
in every part of nature
that is perfectly regular
and geometrical if
we can but find it out.*

JOHN WOODWARD,
1699

2. Leaves From Naturalists' Notebooks

For many centuries, and around the world, man has observed and marveled at the periodic rhythms of the living things around him. These observations extend from Aristotle's (384–322 B.C.) study of sea urchins to the present day.[1] Remarkable reports of the time sense of plants and animals have come from Greece, France, Germany, Switzerland, the West Indies, Panama, California, the South Pacific, Australia, and Japan.

The effects are so striking, and their timing is so amazingly precise, that those who first recognized them were content to

---

[1] For example, in the spring of 1970, Dr. Arthur L. Jensen, the noted dental authority, told me that many of his patients are sensitive in regular cycles. At one phase of the cycle a patient can accept dental work readily; at the opposite phase he is unable to bear it. Dr. Jensen therefore schedules appointments in phase with these rhythms.

report what they had seen. Some of the writers, indeed, were perceptive enough to see that the rhythms were in phase with some natural rhythm of the environment: in phase with the earth's rotation, or the moon's orbit around the earth, or the earth's orbit around the sun. And some even counted the numbers of animals that did certain things at certain times. But none of them did experiments in the modern sense; that is, they did not propose any explanations for the behavior they saw, and of course they carried out no tests. They were content just to wonder at the "procedure of nature that is perfectly regular and geometrical."

And the implications of this strange miscellany went long unrecognized. But when the relationships among these isolated oddities began to emerge, they were seen to be profoundly significant; each case was seen to be a special illustration of a pervasive feature of nature: that virtually all living organisms carry out their functions in rhythmic cycles. From this springboard the experimental biologists who followed projected a whole new science.

For perspective let us follow some of the earlier naturalists, as they observe *how* animals carry out these rhythmic functions. With this background we may then ask: *Why* do animals do this? And *by what means?*

One of the earliest observers of what may have been a rhythmic display did not recognize it as such at the time, and more than five centuries passed before a modern marine biologist suggested that possibility. The observer was Christopher Columbus.

It was ten o'clock on the moonless night of October 11, 1492. Four bells clanged by the first watch. Captain General Christopher Columbus braced his legs on the careening poop castle as the decked ship *Santa Maria* scudded into the southwest, a subtropical gale lashing at their backs. The near-mutinous crew grumbled in the ship below. Time had nearly run out since Columbus had stood alone before his rebellious

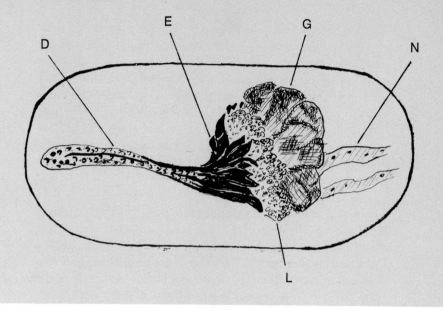

The light gland from which the Bermuda fireworm emits its brilliantly luminous secretion at mating time. Most of the many body segments of polychaete sea worms are equipped with two swimming paddles, or parapodia, one on each side of a segment. This section of one such paddle shows the light gland. From left to right, (D) is the duct of the gland; at (E) are accessory cells of the light gland; at (L) are light cells with granular masses (G). The luminous secretion is discharged from the excretory organ, or nephridium, (N).

men; he had promised to turn back if three days did not bring landfall.[2]

Hints of lush lands beyond the horizon had lent hope in the anxious days before. Bushes laden with aromatic berries had drifted by on the indigo sea. Land birds had whistled high overhead and faded into the southwest. A hand-carved stick was hauled aboard, dripping. Still the crew muttered darkly.

[2] Albert B. Donworth: *Why Columbus Sailed* (New York: Exposition Press; 1953), pp. 131–3. Samuel Eliot Morison: *Christopher Columbus, Mariner* (Boston: Little, Brown and Company; 1955), pp. 47–8.

Thirty-two days had passed since the Canary Islands had sunk into the eastern sea. Gale-driven, *Santa Maria* pitched and rolled, rushing under full sail into the darkness.

Peering into the moonless void, Columbus suddenly saw an eerie light ahead. Was it land? According to the turbulent Spanish historian Bartolomé de las Casas, the first to see land that night was Rodrigo de Triana, a sailor aboard *Pinta*. For our purposes it makes no difference who saw it first. The important point—generally agreed on—is that the Admiral did see a flickering light.

In his précis of *Historia de las Indias* las Casas wrote: "As the caravel *Pinta* was faster than the other two and ahead of the Admiral, she discovered the land and gave the signals that the Admiral had ordered. He who first saw the land was a sailor called Rodrigo de Triana, notwithstanding that the Admiral, at ten o'clock that evening, while on the castle of the poop had seen a light. But the thing was so indistinct that he did not dare assert that it was land; nonetheless, he called Pedro Gutierrez, confidant of the King, and told him that he seemed to have seen a light, and that he should look too, which he did, and he saw it. . . . After the Admiral had spoken, this light was seen once or twice more; and it was like a bad waxen candle that rose up and went down, which few would have thought an indication of land; nonetheless the Admiral was certain of being close to it."[3]

What could that mysterious light have been? An illusion? Samuel Eliot Morison, authority on Columbus and the lore of the sea, thought so. "Volumes have been written to explain what this light was or might have been. To a seaman it requires no explanation. It was an illusion, created by tense watchfulness. When uncertain of your exact position, and straining to make a night landfall, you are apt to see imaginary lights and flashes and to hear nonexistent bells and breakers."[4]

[3] Bartolomé de las Casas: *Historia de las Indias*, (1527–61) quoted in Jean Merrien (pseud.): *Christopher Columbus, the Mariner and the Man* (London: Odham's Press, Ltd.; 1958), pp. 146–7.
[4] Morison: *Columbus*, p. 48. Morison continued this view for many years after an explanation based on biological facts had been published.

Across the Atlantic, at the mouth of the Tamar River, Plymouth Sound laps the base of ancient stone walls rising in terraces toward a hilltop. Beyond broad lawns and close-cropped English hedges stands a white fortress of a building—the Plymouth Laboratory of the Biological Association of the United Kingdom.

In his laboratory one day in the early 1930's, marine biologist L. R. Crawshay studied his notebook, his mind teeming with memories of the ocean life in the Bahamas, where he had recently been stationed. He recalled the brilliantly luminous displays of the Atlantic fireworm that he had studied there. This annelid marine worm had fascinated him because it timed its nocturnal sex activities by the phases of the moon. Groups of six to twenty females of the genus *Odontosyllis* would appear suddenly at the top of the water, shedding eggs and discharging streams of brightly luminous secretion. Then

The commonly accepted historical view has been that Columbus's first landfall was San Salvador. More recent studies of the precise lunar mating rhythms of the Bermuda fireworm suggest that this first landfall may have been Cat Island.

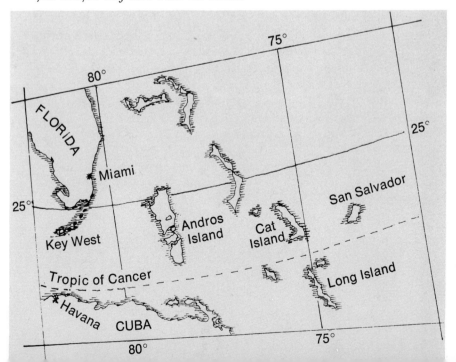

the males would flash like fireflies and rush in to fertilize the eggs. The fireworks would last for no more than ten minutes in all.

Crawshay had carefully pinpointed the timing of these displays. They always took place during a particular eighteen-hour period in each lunar month; and they centered around a time just one hour before moonrise—on the night before the fourth quarter of the moon.

Now, back in Plymouth, Crawshay made an imaginative association of two known facts. Columbus, approaching the Bahamas, had seen a light that "waxed and waned," but he had never known its source. The fireworm, *Odontosyllis,* native to the Bahamas, displayed just such a light—at a particular time of night at a particular phase of the moon. Could that have been Columbus's flickering light?

Carefully checking old almanacs, Crawshay found that the night of October 11, 1492, had been a night just before the fourth quarter of the moon. The moon had risen just one hour after Columbus had seen the light. Could such a coincidence have occurred through pure chance? The probability was small indeed. At the very least Crawshay had had the first novel idea about the famous light in nearly five hundred years, and he published it in a paper that he wrote for *Nature* in 1935.[5]

In the same paper he discussed the favored haunts of the fireworm in relation to sea mileages and ocean depths off the Bahamas, and suggested that the first landfall may have been Cat Island, rather than San Salvador, as most historians have it. I doubt if the truth will ever be known, but at any rate Crawshay's speculations were based on observed and verifiable facts—not on old documents.

Recognition of the effect of the moon on marine animals is not new with the twentieth century. Aristotle wrote of the swelling of the ovaries of sea urchins at full moon; he even

[5] L. R. Crawshay: "Possible Bearing of a Luminous Syllid on the Question of the Landing of Columbus," *Nature* (London), Vol. CXXXVI (1935), pp. 559–60.

The palolo worm of the South Pacific shown magnified three times. Svante Arrhenius proposed that the reproductive cycle of the palolo was determined by variations in the electrical charge of the atmosphere.

described these spiny creatures in such detail that their chewing organ is known to zoologists today as Aristotle's lantern. Cicero said oysters and other shellfish increase and decrease in number with phases of the moon, and Pliny confirmed this. The last quarter of the moon is especially well suited to the sexual activities of sea creatures everywhere.

Halfway around the world from San Salvador, and fifteen degrees below the equator, lives a polychaete sea worm that the natives of Fiji and Samoa call the "palolo." This ocean denizen grows up to eighteen inches long and lives in dim caverns in the coral reefs. Its main sex cycle occurs just once each year, during the last quarter of the moon in November—spring in the South Pacific. Millions of individual worms "know" when the magic moment has come.

Before the November swarming, the hind part of the palolo swells with sex cells. At dawn this hind part breaks off, surges to the surface and thrashes about, gradually breaking into pieces to free the eggs and sperm. The tropic ocean simmers like vermicelli soup—a greenish brown.

Ringing the Koro Sea, Fiji's three hundred islands endure the sweltering sun and fitful rains of the tropics. Tiny Ovalau Island, steeped in the history of the early capital town of Levuka, maintains the residence of the commissioner. One evening early in the 1940's Commissioner William Burrows, warm but immaculate in white linen, crossed the room for a breath of the eastern trade. Raindrops rattled on broad banana leaves in the garden below.

The commissioner had been thinking how curious it was that the palolo should swarm at the same season each year, and he wondered whether earlier accounts had recognized just how precise the timing was. He recalled that Basil Thomson had written of the palolo some fifty years before, and he turned to *The Confessions of Lady Asanath*.

"The time of the annual swarming is a great occasion for the natives, as the palolo is regarded as a delicacy. . . . Cakes of the gelatinous mass are fried, and taste just like oysters." Basil Thomson, then, had said only that the swarming was annual.

Since Commissioner Burrows was in close touch with local lore, he knew that one of the old natives was said to forecast the swarming to the day. Such precision shown by a sea worm should be of general interest, Burrows thought, and he wrote his impressions of the whole exotic performance for London's *Nature* magazine, where he told of the "unforgettable event" in colorful detail:

> During my period as commissioner, stationed at Levuka, I have seen the "rising" at Tokou on several occasions. Nearly

every year there are two risings, the first known as *Mblolo lailai* (small Mblolo) and the second as *Mblolo levu* (large Mblolo); "large" and "small" do not refer to the size of the worm but to the quantity of the worms. Occasionally there is no *Mblolo lailai.*

In a normal year the *Mblolo lailai* appears about the end of October and is followed by the second, and main rising about two or three weeks later. The main rising always occurs at dawn, and literally, the worm comes up with the sun. It is, also, always at the time of high water.

The worm, when it comes to the surface, is headless, and it is known that the head remains alive in the reef. The parts shed are from ten to fifteen inches long when they reach the top of the water and continue to wriggle. There is an old Fijian living in the village of Tokou who was a foreteller of the day of the rising. . . .

When the first light of dawn appears, great funnels of worms burst to the surface and spread out until the whole area is a wriggling mass of them, brown and green in color.

When the tropical sun rises perpendicularly from the sea the catch is in full swing, and hundreds of boats, canoes and punts are filling up kerosene tins and jars by the simple process of dipping them out with nets. The worms also provide an annual feast for the fish; for all around and between the boats big fish and sharks cruise quietly along, gulping them in, and take no notice whatever of the boats or their occupants.

As the sun makes itself felt, a change begins to occur in the length of the worms. They begin to break up into shorter and shorter bits, until some three hours after sunrise the entire surface of the sea shows nothing more than patches of scum. . . .[6]

The commissioner laid down his pen. Perhaps a gin *pahit,* and then to bed. He'd have his letter got off in the morning.

"Grunion run tonight!"
Visitors to Southern California beaches listen skeptically.

[6] William Burrows: "Periodic Spawning of 'Palolo' Worms in Pacific Waters," *Nature* (London), Vol. CLV (1945), pp. 47–8.

The grunion of Southern California spawns at the most favorable period in the tidal cycle for survival of its young. The female deposits her eggs in a hole she bores into the sand with her tail. The males curve around her and release sperm to fertilize the eggs.

They have heard the tales of grunion runs, and tend to suspect them at first hearing, like snipe hunts and similar folk humor. But grunion runs are real; more obvious displays of the rhythms of nature are seldom seen.

On warm summer nights from Point Conception to Punta Abreojos—Santa Barbara to Ensenada in Baja California—thousands of people line the shore in wait for grunion. It is a day or so after the moon has reached its full. Shoes and stockings are dropped on the sand; pants legs are rolled up and skirts gathered. Everyone has a bucket, a glass jar, or a cloth sack to fill, for the fish may be caught legally only with bare hands.

The incoming waves rumble shoreward; the tide is one day past its first monthly high. In fifteen minutes, the grunion run will start. As the tide first recedes, thousands on thousands of slender silver fish surge to the crests of the breakers and wash onto the moist sand. The beach shimmers like a tinsel sea. The reproductive cycle begins.

The female grunion is the length of a man's hand. Dancing

upright on her tail, she digs herself down into the sand for half the length of her body. Swaying wildly, she deposits her eggs in the cavity three inches below the surface. The male wriggles up and curves his body in a graceful arc around her as she sways. In a few seconds he has fertilized the eggs and is slithering back toward the waves. The female, now utterly exhausted, sways and wriggles to free herself. Then she flips toward the water and vanishes into the breakers; the wet sand slides into the hole and covers the eggs.

Meanwhile the eager grunion hunters are scooping up the squirming fish and tossing handfuls into sacks and buckets. Children throw themselves down, screaming and wriggling like the fish themselves, in mad pursuit of the quarry.

Ocean-fresh and broiled sizzling over a driftwood fire, the grunion is a mouth-watering treat; hungry men have been known to down twenty. After the feast, the accuracy of the forecast makes quiet talk. Guesses are chanced on whether tomorrow night will see a second run, or if the slender fish will disappear for another fifteen days. Then the night chill settles on the beaches; the driftwood fires are smothered, and the fishermen go home.

The grunion times its nuptial night so that its fingerlings will have the best odds for survival. The highest tides of the month come during the full and dark of the moon. The grunion chooses a night just a day or so later than these, when high tide will be just inches lower. The eggs, buried snugly in the sand above high water, mature and are ready to hatch in fifteen days—just as the next high tide of the month arrives. In the churning waves and sand of that high tide, the baby fish are jostled free of the orange-colored egg pods to join their parents in the ocean.

The lady from the Middle West is still not convinced. "I think these Southern California newspapers are pulling our leg. This afternoon they said, 'Grunion run expected at 10:37 tonight.' I'll bet that was just a wild guess."

A balding man in damp coveralls massages white stubble on his jaw. "Well, they *did* come in at 10:40. That's close enough. Really, though, all you need is a tide table. The runs always come just fifteen minutes after high tide—that is, after the two highest tides of the month."

So it is quite easy to predict when the grunion will come to the beach to spawn. But it is hard indeed to tell how they "know" so accurately the most favorable hour for the later survival of their young. Scientists have suggested two entirely different mechanisms to explain such rhythms as depend on the phases of the moon. One of these is related to daily rhythms, and the other assumes that the organisms are highly sensitive to variations in the character or intensity of the light that falls on them.

To appreciate fully the implications of these two views, we shall first need to see more of the experimental work that biologists have done on daily rhythms and on the effects of light on rhythms of other lengths. Still, the general outline may be foreshadowed.

Erwin Bünning, whose work we shall look at in more detail later on, has pointed out that lunar or semilunar rhythms appear at first glance to have no relationship to daily rhythms, but that actually there is evidence that these rhythms may result from tidal rhythms that are reinforced by daily rhythms arising within the organism.[7] The other hypothesis is that it is the light of the moon that stimulates sea creatures to sex activity. The crest of the grunion run does indeed occur near full moon. In contrast, the *Odontosyllis* that Columbus probably saw is active before moonrise, so that exposure to moonlight cannot explain both phenomena. Furthermore, even when the moon is full, its light is only as bright as the sun's by one part in half a million, and Cloudsley-Thompson finds it hard to see how such a low light intensity could be responsible for the lunar reproduction cycles.[8] Other biologists have done experiments that indicate that the total time of illumination may be more critical than the intensity of the light.

---

[7] Erwin Bünning: *The Physiological Clock: Endogenous Diurnal Rhythms and Biological Chronometry* (New York: Academic Press, Inc.; 1964), p. 104.

[8] Cloudsley-Thompson: *Rhythmic Activity in Animal Physiology and Behaviour* (New York: Academic Press, Inc.; 1961), p. 90.

We shall look at this work in more detail in later chapters, but for the moment let us review some early observations of the effects of large variations in light intensity that occurred at unusual times of the day. Such large variations would obviously occur during eclipses of the sun.

An English surgeon named George Newport knew that an eclipse of the sun would be seen over England on May 15, 1836. He kept two hives of bees at Chichester, Sussex, and since he was a member of the Entomological Society of London as well as a member of the College of Surgeons, he thought it would be interesting to see how the bees behaved during the eclipse. He made a careful record of all of his observations and sent a full account of them to the Royal Society of London for publication in the *Philosophical Transactions*.[9] On the following page is one set of his notes.

Thus it is obvious, even from a record nearly a century and a half old, that the activities of some animals are critically influenced when the light intensity is reduced—even though that reduction occurs at an abnormal time of day. This in turn suggests that the timing mechanism can be reset by changing the normal cycle of light and dark, and we shall see later how important this effect is in present-day research on living clocks.

A hundred years after Newport, the Japanese biologist Syuiti Mori decided to make a more thorough series of observations of animals during the eclipse that was to take place over Hokkaido on June 19, 1936. By then, of course, the concepts on which he could plan his work were more advanced, and the techniques he could use were more sophisticated than anything that had been available to Newport.

For some of his tests, Syuiti Mori chose a fly of the genus *Protofucellia* because it is normally active only in the daytime. How would it react during a solar eclipse? Mori designed a test rig of elegant simplicity. He planned to lure

[9] The Royal Society of 1837 gave nearly eighty pages to this record. While no scientific society today would allow observations like these more than a few paragraphs, it was just such detailed work that had to precede the great inductive concepts like Darwin's on evolution.

TIME, P.M.	REMARKS
1:30	Abundance of bees around the hives. Loud humming.
2:00	Eclipse has commenced, many drones abroad, bees greatly excited, flying around hives.
2:15	Sunlight sensibly diminished, bees flocking home, very few go abroad.
2:30	Light still diminishing, scarcely a bee goes abroad.
2:45	Bees flocking home very rapidly, a few drones still abroad.
3:00	Light greatly diminished. *Geotrupes stercorarius* [a beetle that flies at twilight] on the wing.
3:15	Light more obscured, hives quiet as in the evening, not a bee goes abroad; cocks crowing, town in the distance very hazy, cool light wind, sky very clear.
3:20	Eclipse past its maximum, two bees have just come home again.
3:30	Light sensibly increased, bees at the entrance of the hives and going abroad.
3:45	Light still increasing, a few bees going abroad.
3:50	Light much increased, bees still going abroad.
4:00	Great increase of light; one bee has again returned with pollen.
4:20	Eclipse nearly terminated. But few bees abroad from hive No. 1.
4:30	Bees abroad from hive No. 2; eclipse terminated.
5:00	Bees abroad from both hives; sky clear, very fine.[1]

[1] George Newport: "On the Temperature of Insects and its Connection with the Function of Respiration and Circulation in this Class of Invertebrate Animals," *Philosophical Transactions* (1837), Table VII, p. 307.

the flies with the fetid smell of fish viscera placed in a tin can, and to count the number that he could catch on a sticky ribbon an inch and a half wide and fourteen inches long.[2]

"The flies soon gathered around the can," Mori wrote, "and consequently some individuals stuck to the trap. After 30 seconds the sticky ribbon was taken off and the number of flies on it was counted and recorded. For each experiment the sticky ribbon was renewed. The number thus obtained was taken as the activity of the animal."

First Mori had to find out how the flies behaved under normal conditions—the control test. On June 16, three days before the eclipse, he made continuous counts with his fly-trap. An hour before sunrise, he caught no flies; as the run rose, he caught 48; in the half-minute before eleven in the morning, he caught 72. By sunset he caught only 9, and after dark, again none. The flies had quit for the day.

As Mori summarized it: "The fly was active only during the daytime. We can assume its closer correlation with the intensity of solar illumination than with any other environmental factors such as air temperature or relative humidity. This is clearly manifested by the calculation of the coefficient of correlation between the number of insects and the change in environmental conditions."

On the day of the eclipse, the fly counts were similar to the control day—up to the time the eclipse began. Then, just as the moon's first shadow appeared at 2:08 in the afternoon, Mori caught only 55 flies. By 3:20 it was completely dark, and the fly count dropped to 14. After full daylight returned, the number jumped back to 48.

"As the eclipse advanced and the intensity of illumination decreased the activity of the fly fell off and the number of flies caught in the trap diminished. The time of the minimum activity coincided perfectly with the time of the total eclipse.

[2] This summary is based on Mori's delightfully informal paper, "Effects of the Total Solar Eclipse on the Rhythmic Diurnal Activities of Some Animals," *Annotationes Zoologicae Japonensis*, Vol. XVIII (1939), pp. 115–32.

Then as the sunlight grew stronger and stronger, the fly too gradually recovered its activity.

". . . we recognize that the diurnal rhythmic activity of the fly, so far as its behavior in flocking to food is concerned, belongs to a dependent type of rhythmic activity, regulated chiefly by the diurnal march of the sunlight, but partly by air temperature and consequently by relative humidity. One reason for the slight discordance between the value of the coefficient of correlation for the sunlight on ordinary days and on the day of solar eclipse may be due to too rapid change of the light during the eclipse. If the totality had had a more prolonged duration, activity of the fly might have been more reduced."

Altogether it was a busy day for Syuiti Mori. Between fly counts he also made notes on "the roosting of crows and the crowing of roosters," as well as on the swarming of midges and the croaking of tree frogs. All of these animals cut their activities as the unnatural darkness came on.

". . . these facts appear to indicate that diurnal activities on ordinary days belong to the class of dependent rhythm, not to the independent hereditary type. The solar eclipse is a great experiment conducted by nature creating out-of-time modification of environmental conditions in a large scale. It is a good opportunity to study such types of behavior."

Still, there is a strange difference here, Mori would remind himself. "I watched sandhoppers, and starlings, and mynah birds too; but they paid no attention to the eclipse at all. They were just as active as though the sun were shining brightly. Crayfish too. The lovable school-children had gathered them for me. The pigment in the eye of a crayfish migrates at night; but the darkness of the eclipse had no effect on this. Why should this be so?"

Syuiti Mori was not the only biologist to run into such puzzling differences.

Wedged between Saudi Arabia and Iran, the southern tip of Iraq fronts the Persian Gulf for a scant few dozen miles.

Nevertheless the people of southern Iraq have an unreasoning fear of the whales that may lurk offshore.

On June 25, 1952, the biologist N. A. Weber waited in southern Iraq to study the effect of a coming solar eclipse on every animal he could keep track of.[3] As the eclipse began, the air vibrated to the rattle of drums and the clatter of hammered pots and pans. Keening above the uproar rose terrified wails from the natives.

A smile of understanding crossed Weber's face. No matter that the natives were victims of an ancient superstition. Their terror was real. In an eclipse a whale tries to swallow the sun; if it succeeds, the sun will be lost forever. To avert that cosmic tragedy, the natives raised the loudest uproar their fright could muster—to frighten the whale away.

After the eclipse was over, Weber summarized his bulky notes. During the darkness near total eclipse, cockroaches came out, just as they do in the normal dark of night. Crickets set up a nocturnal chorus of chirping. Some of the mammals appeared to be affected, but others were not. Most birds showed responses that Weber could classify as modified, and the flies, moths, and honeybees sought cover. But many animals, so far as Weber could tell, were not influenced at all.

What could cause the difference? Why should some animals react to unseasonable darkness, while others were unaffected? To start from fundamentals (Weber reasoned), all of the animals—under normal conditions—had rhythms that were timed to the earth's rotation. Nocturnal animals began stirring at dusk, and diurnal animals woke near dawn. Weber knew that there were two schools of thought about how these rhythms were timed; one view was that they were timed by some clocklike mechanism within the animal—a purely internal rhythm. The other view was that the animal was responding to an external influence. So, Weber thought, a simplistic explanation would be that the animals that were influenced by the eclipse were timed externally, and those

[3] N. A. Weber: "The Animal Behaviour Eclipse Expedition of the College of Arts and Science," *Baghdad College of Arts and Science Publications,* No. 2 (1952), pp. 1–23.

that were not influenced were timed internally. And that, Weber thought unhappily, really got nowhere.

Perhaps a better clue to the real cause might be found in the way bees time their foraging activities. Weber was familiar with a striking observation that had been made by the Swiss psychiatrist Auguste Forel as long ago as 1906.

On fine mornings near the beginning of this century Forel used to eat breakfast with his family on the terrace in his garden. Jams and jellies freshly made from local fruits were always on the table. Very soon, honeybees arrived to help themselves. Dr. Forel quickly noticed that the bees came punctually at breakfast time, and he began to time them with his watch.

Some days, breakfast was served in the house—not on the terrace. One of these mornings, Forel saw something that puzzled him. The bees were on the terrace at their usual time even though no jam was there. He had supposed that the bees were attracted simply by the fruity aroma, but they came to the terrace every morning on the dot, jam or no jam. From his boyhood, Forel had spent long hours studying the behavior of ants, and he checked over everything that he could remember about ant societies to see if he could find a clue. But he found none. He had a vague idea that the bees were somehow able to tell time, but he made no effort to learn anything further about it.

And Weber's dilemma remained.

A few years after Forel, the German scientist von Buttel-Reepen took much pleasure in strolling through richly colored buckwheat fields. The fragrance of the flowers, the warmth of the midmorning sun, and the humming of the bees collecting nectar, gave him a languorous sense of well-being. But a puzzle kept demanding an answer. The buckwheat blossoms stopped their flow of nectar in the middle of the morning. The bees promptly left and stayed away until the next nectar flow was due to start.[4]

[4] Max Renner: "The Clock of the Bees," *Natural History*, Vol. LXVIII (October 1959), pp. 434–40.

Pondering, von Buttel-Reepen would go back to the buck-wheat fields on sunny afternoons. Bending over a flower he would draw in a deep breath. It seemed to him that the flowers were just as fragrant in the afternoon as they were in the morning. But the bees never came in the afternoon. Clearly it was not the fragrance that drew them. Then how *did* they know when to come?

Von Buttel-Reepen made a guess about why. He did not clearly realize it, but his guess was only one of at least three possible explanations he could have chosen.

—He could have supposed that the bee colonies kept a few lone scouts watching the fields. When the nectar flow started, the scouts might fly back to the hive and report.

—He could have thought the bees related nectar flow to some external event—the position of the sun in the sky, for example. If they remembered when the flow started yesterday, they could forage when the sun was in the same position today.

—He could have guessed that the bees had an actual time sense—an internal clock. That was the guess he made, and it was a good one.

To describe the idea, he coined the word *Zeitgedächtnis* (time sense), a term so apt that it is now found in biological literature in English almost as often as in German. Was the guess a lucky one? Probably so, because it was out of line with a working rule that has guided scientific reasoning for many centuries—the principle of parsimony.

"Science is really the search for simplicity. William of Occam, a fourteenth-century philosopher, made the dictum . . . 'Entities should not be multiplied beyond necessity.' This principle of parsimony (often called *Occam's razor* because it pares a theory to its bare essentials) means that no more forces or causes should be postulated than are necessary to account for the phenomenon observed. In practice, this means that the simplest explanation which will account satisfactorily for all the known facts is to be preferred. A new theory in biology, by clearing away previous misconceptions and by pointing up new interrelations of phenomena, not only stimulates research in theoretical biology, it also provides a basis for a host of advances in medicine, agriculture, and similar fields."[5]

If von Buttel-Reepen had assumed that scouts reported back to the beehives, or even that the bees started foraging when the sun was in a favorable position, he would have hit on a simpler explanation than his concept of a time sense. But perhaps he did not realize that simpler explanations were possible. In any event, he had no way of proving his case. In his day, and in Forel's, precision methods in biology were barely in their infancy. It was not until 1929 that a woman student of the great Karl von Frisch gained deeper insight into the mystery of the *Zeitgedächtnis* in bees, as we shall see later.

Man has been fascinated by the ways of birds ever since Noah's dove returned with the olive leaf. According to popular accounts, many birds display clocklike regularity.

One bird species—resembling a tiny ostrich—is so punctual that at home in Panama it is called the three-hour bird. This pileated tinamou sings every three hours, day or night. According to the Panamanians, one can set his watch by the song of the three-hour bird, and it will check with the airport clock within a few minutes.[6] They do not claim to know why the song is so punctual; but they do report that it is the male that incubates the eggs, and takes care of the chicks, and bursts into song eight equally spaced times in each calendar day.

Note that the tinamou sings both day and night. If the bird gets some clue to the time from some outside source, it is certainly not from the sun. An inherited internal clock would explain it. So would a compound rhythm of subtle cosmic forces, but this would require a more complicated theory than Occam's razor would allow.

The naturalist Frank E. Lutz listened day after day for the wren's first song in the morning. "It is amazingly precise,"

[5] Claude A. Villee: *Biology* (Philadelphia: W. B. Saunders Company; 1967), p. 4.
[6] This and several other popular observations of animal behavior that follow are based on Frank W. Lane: *Animal Wonder World* (New York: Sheridan House; 1951), pp. 141–62.

he reported. "Morning after morning I heard the song to the minute on the clock, between 5:57 and 5:58 A.M. This is indeed a puzzle when you remember that the sun changes its position by four or five minutes every day."

In the Luxembourg Gardens in Paris, a flock of sparrows flew within minutes of 9:45 each morning, waiting for a visitor who came regularly to feed them. Then came a day when the time changed to daylight saving. Did the birds remain on standard time and arrive an hour late? It is hard to believe, but it was said in Paris that the birds adjusted to daylight saving time overnight!

W. H. Hudson, of *Green Mansions* fame, has told of a wild teal that adopted an English country family and became attached to the head of the house, who commuted regularly to the city. No matter where the little fresh-water duck had been during the day, she would always appear at the front of the house a few minutes before her master was due to return. If he was late, she would wait patiently for him for at least an hour, her eyes glued to the path down which he regularly came.

The northwestern flicker has the curious habit of hanging itself up to sleep. J. Hooper Bowles says that he was once spending an afternoon with a friend and happened to remark that it was half past three. He was surprised to hear his host say, "Then in five minutes it will be bedtime for our flicker." They went outside and watched. Within five minutes the flicker appeared and hung himself against a board under the eaves, where he spent the night. The bird had been doing this with perfect regularity for some time. Exactly at 3:35 each afternoon, even though it was broad daylight and bright sunshiny weather, this woodpecker always called it a day and went to bed.

A great many of the periodic activities of animals are related to the reproductive cycle. In birds, the courtship displays are often elaborate. The wild spruce goose performs to a timed rhythm. J. L. Devaney says the gander gyrates through a long series of weird antics to woo his mate: "[His] favorite location at such a time is between two trees standing some twenty or thirty feet apart, and with the lower branches large and horizontal. Perched on one of these branches he pitches downward, pausing midway to beat and flutter his wings, and ascends to a branch of the opposite tree. After a short interval, his maneuver is repeated and so continued by the hour, the bird swinging back and forth from tree to tree, the time between each swing being as exact as if measured by a watch."

Birds are also clock-watchers about feeding. Gregarious curlews band together to feed on mudflats and marshes. Tall and stately, they search through the ooze with down-curved bills for morsels of animal food. When the tide rises and covers the mud flats, the curlews climb on strong wings and fly inland to feeding grounds that are actually less to their liking. Though large birds, their gray-streaked brownish feathers blend into the dun background of the fields. Feeding rather unhappily on dry land, they wait impatiently for the tide to uncover the mudflats so that they can return to their favorite food.

Do they send out scouts to watch the tide level? Some naturalists have suspected that they do, but more careful checking has shown that they do not. Nevertheless, when the tide has gone out enough to uncover the mudflats, the curlews gather together in a tight bunch, rise swiftly, and return to jab their sickle bills into the ooze.

Above birds on the evolutionary scale is the curious Australian flying fox. This nocturnal fruit bat spends its days hanging upside down among close-packed tree leaves; at dusk it flies out to feed. Rising on gray-brown furry wings of five-foot spread, its flight is swift and powerful. "Companies" descend on village fruit plantations to gorge themselves. Unlike many bats, with their ugly pushed-in faces, dog teeth, and beady eyes, the fox bat looks rather like a small fox, with whiskered muzzle, arched neck, and big liquid eyes. Also somewhere within his body is a well-regulated clock.

When these foxes fly out to feed, the Australians call it "breaking camp," and they claim they can set their watches by the flight. But they might wish the affair were not so unalterable; the twilight hordes lay waste to the crops.

Everyone has heard of dogs with a time sense. People raised on farms know dogs that regularly go out at the right time to bring the cows home for milking. In one case, the dog quickly adjusted to daylight saving time.

In another case, it was the cows that showed the time sense. One summer day in 1888, a physician named A. S. Hudson impatiently put down his copy of *Popular Science Monthly,* annoyed by an article that ridiculed the idea that animals have a time sense. An experience of Hudson's own had convinced him that they do, and he wrote a letter to the editor to set the record straight:

> Something like a half a century ago, the writer had the care and milking of five cows during one summer. They grazed in a pasture lot many rods from the dwelling. It was the custom to give the animals salt every Sunday morning. They enjoyed the treat, and it was evident that they began to expect it. After a length of time—I cannot say how long—a curious behavior of the cattle became conspicuous, for every Sunday morning they were found standing at the bars, the point nearest the house, with every appearance of mute expectation. At every other morning, as well as at evening, they had to be sought and brought to the bars for milking. Sometimes I would forget to take the salt with me at the stated time, when, instead of moving off to feed when my task was done, as they usually did, they remained about the spot for an hour or so, as if waiting for their weekly ration of salt.

> Here, then, is the problem: every Sunday morning these cows came of their own option to the place of milking, where on that day they generally got salt, and not on other mornings. How could they do that except through some faculty of estimating the seven days of the week? . . . They were isolated from the outer world, away from any thoroughfare, and saw no one but myself from one week to another and from one month to another. So far as we could judge, one day was like all days except Sunday, which they might have called *salt-day,* had they possessed the faculty of speech. How did they note that cycle of time, to be there on that morning and not on any other morning?[7]

[7] A. S. Hudson: "Can Animals Count the Days?" *Popular Science Monthly* (November 1888), p. 123.

A chacma, or Cape baboon, who put in twenty years on exhibition in the U. S. National Zoological Park in Washington, D. C., set his own quitting time—and observed it to the minute. Promptly at four every afternoon he scratched his rough coat of sooty green, lumbered into his private house, and shut the door. If late visitors wanted to see him, they could come back the next day.

Too, many people have remarkably acute senses of time. H. Mortimer Batten knew a gypsy boy who had never owned a watch, but always kept appointments on the dot. He could tell the time of day within a few minutes of the clock. Asked how he did it, he laughed and said: "Just know."

These are but a few of the leaves from naturalists' note-books that have aroused the curiosity of twentieth-century biologists and spurred them to devise new concepts and test them by such advanced techniques as computerized statistical analysis, radar study of bird and insect navigation, and scanning electron microscope scrutiny of living specimens invisible to the naked eye in the lifelike three dimensions of its viewing screen. Many of these early observations were made by men who were not even trained in the science of their day, yet Colin S. Pittendrigh, a current world authority on biological rhythms, has said: "I can think of no other case in which purely natural historical observation leads so directly—when fully evaluated—to the formulation of first rate problems for the [modern] cell physiologist."[8]

In later chapters we shall watch cell physiologists wrestling with some of these problems. For perspective, however, we shall first review the transition in thought bridging the earlier period of observation and speculation and the current period of sophistication in the tactics and strategy of science. That story begins inside a French astronomer's closet.

---

[8] Colin S. Pittendrigh: "Perspectives in the Study of Biological Clocks," *Symposium on Perspectives in Marine Biology* (Berkeley: University of California Press; 1958), p. 239.

3. Transition: De Mairan to Arrhenius

We who have seen men walk on the surface of the moon will find it hard to narrow our view of the world so as to feel rapport with the scientists of the eighteenth century. But for perspective on the story of living clocks, let us try.

Picture the year 1729 in Paris. A visitor would soon be lost; hardly a street sign was to be seen, and the few that had been put up were carved in stone set into the walls of corner buildings.

For the scientists bent on exploring nature there were even fewer reliable signs. Newton's universal law of gravitation had been announced only a few decades earlier, and Pierre Laplace, "the Newton of France," was yet to be born. The phlogiston theory (maintaining that fire is a material substance) lay in the path of chemistry like a giant stumbling block. A quarter of a

The sensitive plant of M. de Mairan awakes at dawn, even though it has been kept in total darkness. The inset at lower left shows the position of the leaves when the plant sleeps.

century was to pass before Benjamin Franklin drew static electricity from a kite string and showed that it was only a natural phenomenon, and not some terrifying display of the anger of the gods. Laboratory apparatus was crude; in medicine, even the stethoscope lay nearly a century in the future. Nevertheless, by 1729 the Age of Enlightenment had progressed far enough that scientists were eager to apply the experimental method to the endless array of questions that lay before them. Such is the setting in which we find the French astronomer Jean Jacques d'Ortous de Mairan and his friend Marchant, arguing about the publication of a paper in the Proceedings of the Royal Academy of Paris.

Marchant was insisting that de Mairan write a report to the French Academy describing de Mairan's discovery about the sleep movements of a sensitive plant—a discovery that struck Marchant as truly astonishing. De Mairan refused to take the time. To him, the discovery was no more than an especially curious sidelight. As a busy astronomer, his mind was filled with unanswered questions about the aurora borealis, and the relation of a prism's rainbow colors to the musical scale, and the diurnal rotation of the earth, and the satellites of Venus, and the total eclipse of the sun that had occurred in 1706. He would waste no time writing to the Academy about the sleep of a plant!

But Marchant was equally determined. Here was the first proof that a plant could sense the position of the sun without being exposed to it in any way, and he felt that it was a duty to report the story to France's scientific body. It was quite customary for a member of the Academy—which Marchant was—to report the work of another scientist, and this he did of de Mairan's work, in 1729. In free translation his account follows:

"It is well known that the most sensitive of the heliotropes turns its leaves and branches in the direction of greatest light intensity.[1] This property is common to many other plants, but the heliotrope is peculiar in that it is sensitive to the sun (or time of day) in another way: the leaves and stems fold up when the sun goes down, in just the same way as when one touches or agitates the plant.

"But M. de Mairan observed that this phenomenon was not restricted to the sunset or to the open air; it is only a little less marked when one maintains the plant continually enclosed

---

[1] The name "heliotrope" comes from the Latin, meaning "turning toward the sun." We do not know which species of heliotrope de Mairan used; in fact the first logical system of plant classification was not published until 1735, by Linnaeus. Before that, living things had simply been designated as "fish, birds, beasts, trees, herbs, or shrubs."

in a dark place—it opens very appreciably during the day, and at evening folds up again for the night. This experiment was carried out towards the end of one summer, and well duplicated. The sensitive plant senses the sun without being exposed to it in any way, and is reminiscent of that delicate perception by which invalids in their beds can tell the difference between day and night."[2]

In concluding his letter to the Academy, Marchant invited botanists and physicians to pursue the matter further, while at the same time admitting that "the advance of true natural philosophy, which is experimental, can only be painfully slow."[3]

But no botanist or physician came forward to accept the challenge, and it was nearly thirty years before France produced a mind curious enough to recheck de Mairan's discovery and pursue its implications further.

INSIDE DUHAMEL'S LIGHT-TIGHT LEATHER TRUNK

That mind, strangely enough, belonged to a man who had spent most of his college years in boredom and frustration. During those years, Henri-Louis Duhamel would hardly have been recognized as the man later to be acclaimed "Inspector of the Academy of Sciences and of the Navy, Member of the Royal Society of London, of the Academies of St. Petersburg, Stockholm, Palermo, and Padua; of the Institute of Bologna, the Royal Society of Edinburgh, of the Societies of Agriculture in Paris and Leyden, and Free Associate of the Royal Society of Medicine."

Duhamel was born in Paris in 1700, a descendant of the Dutch nobleman Loth Duhamel, whose son went to France in 1400. While Henri-Louis had the mind of a scientist, it was quite by accident that he stumbled into lectures on natural philosophy, and this only toward the end of his college career.

---

[2] Although the paper was written and presented by Marchant, it is cited in bibliographies as: De Mairan, "Observation botanique," *Histoire de l'Académie Royale des Sciences* (Paris, 1729), p. 35.

[3] On February 3, 1970, while he was leading a genetics seminar at the University of California, Berkeley, Professor Colin S. Pittendrigh was moved to exclaim: "How right he was!"

Then almost overnight his mind caught fire. He listened spellbound as duFay, Lameri, Jussien, Duverney, and Winslow lectured on natural science at the *Jardin du Roi,* and he moved close to the school so that he could give every minute to his studies.

But just as he became totally immersed in his new interest, his parents demanded that he study law. Henri-Louis reacted in a way that was to become characteristic of his method of meeting difficulties.

If he must study law, then very well; he would follow the prescribed course. But he would contrive to pursue his own interests at the same time. In Orléans he could attend a good law school. More to his liking, Orléans was also a vibrant center of industry and the arts, and to these he would give as much time as he wished.

As it turned out, he spared only enough time to the law to earn a degree, and he never used that. After he had thus cavalierly met his obligation to his parents, he returned to Paris full of plans for a life devoted to science, agriculture, and medicine. He studied advanced mathematics, built a laboratory of chemistry, and planned farming tests that would extend for years into the future.

He made careful investigations of the diseases of saffron, of the damage to grains that were caused by the looper moth; and of the causes of the spoilage of flour during shipment. To see if milkwort might have value in treating pleurisy, he brought plants from Virginia to France. He was especially active in urging the conservation of natural resources. For recreation he studied fishing because, he said, "in all its details this art [is] ingenious because it is based on ruse, and very ancient because it provides nourishment."

This was the mind that seethed with curiosity when it heard how strangely the heliotrope had behaved in de Mairan's closet, thirty years earlier.

He asked himself how a plant could possibly move its leaves up and down in daily sleep movements when it had no way of telling whether it was day or night. And he decided that he did not believe it. Perhaps de Mairan had missed a point somewhere. Perhaps his "dark place" had not been absolutely dark. Duhamel determined to repeat the experiment, but only after he had chosen a place where no slightest ray of light could enter.

At once he thought of an old wine cave nearby. It had no air vent through which light might leak, and it could be entered only through a front vault that would serve as a perfect light lock. No ray of sunlight could possibly penetrate to the inner depths.

So it was that one August morning in 1758, Henri-Louis Duhamel carried a spiny sensitive plant through the front vault and into the inner wine cave. But he jostled the plant in his excitement, and it closed up as though it had been struck. Annoyed at his own clumsiness, Duhamel left the plant in the cave and went about his ordinary affairs for the rest of the day. By ten o'clock the next morning, however, he could no longer resist the impulse to look at the plant again, and his candle showed that its leaves were erect and unfolded—it had awakened in total darkness in the cave. And Duhamel's plant continued its sleep-wake cycle for many days, precisely as though it had been directly exposed to the rising and setting of the sun.

Another strange observation puzzled Duhamel as much or even more. Late one day, when the leaves of the plant were still erect, he carried it out of the cave, taking particular care not to jostle it this time. *The plant stayed awake all of the following night!* However, the next evening, the leaves closed again, and thereafter they pursued their normal cycle.

"As the result of this experiment differed from that of M. de Mairan," Duhamel wrote, "I wanted to check whether the difference was due to the fact that the wine cave was darker than M. de Mairan's closet, and for this purpose I put a plant into a large leather trunk covered with heavy wool blankets, and placed in a closed closet. I hoped by this method to keep the plant in total darkness. Nevertheless it opened in the morning and closed in the evening just as in M. de Mairan's experiment."[4]

[4] Henri-Louis Duhamel du Monceau: *La Physique des Arbres* (Paris: H. L. Guerin & L. F. Delatour; 1758), Vol. II, p. 158.

Somehow Duhamel's plant, like de Mairan's—in the complete absence of any light clue—was telling time.

If, therefore, light was not responsible, what other signal might a plant receive from its environment? Duhamel began to wonder whether changes in temperature could be causing the sleep movements. The nights were cooler than the days, and even though he must have realized that the temperature in a wine cave would be fairly constant, Duhamel decided to find out what his plants would do if he heated them well above the normal temperature. His ample means had already provided a hothouse on his own property, and he immediately put some sensitive plants in that. But he soon found that it was not the night chill that sent them to sleep. ". . . I have seen this plant close up every evening in the hothouse even though the heat of the stoves had been much increased," he wrote. "One can conclude from these experiments that the movements of the sensitive plant are dependent neither on the light nor on the heat."[5]

Nearly twenty years before Lavoisier showed that combustion and respiration were simply different rates of the same process, Duhamel had demonstrated that, within limits, a living clock functions independently of temperature—that is, the clock is somehow compensated for the variation with temperature of the internal combustion process. Thus he foreshadowed the concept of a living organism possessing a true chronometer —one that would not run faster or slower as the thermometer rises and falls.

DE CANDOLLE'S MIMOSA CLOCK GAINS TIME

After Duhamel, a century passed before any very important work was done on the sleep movements of plants. The great Swedish naturalist Carolus Linnaeus worked out the first logical system of plant classification, dividing the kingdom into classes, orders, genera, and species. His *Systema Naturae*, first published in 1735, was revised several times before he died in 1778.

[5] *Ibid.*

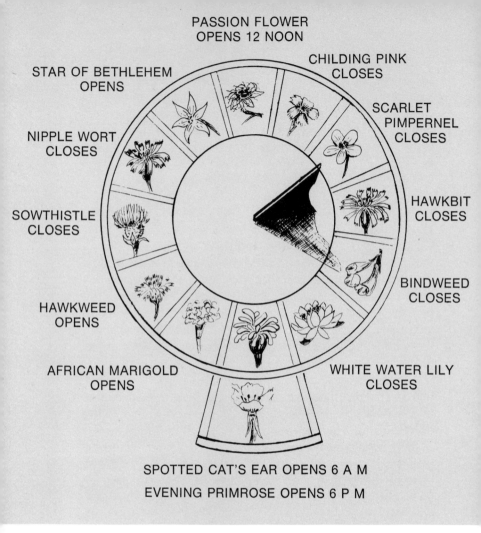

PASSION FLOWER
OPENS 12 NOON

CHILDING PINK
CLOSES

STAR OF BETHLEHEM
OPENS

SCARLET
PIMPERNEL
CLOSES

NIPPLE WORT
CLOSES

HAWKBIT
CLOSES

SOWTHISTLE
CLOSES

BINDWEED
CLOSES

HAWKWEED
OPENS

AFRICAN MARIGOLD
OPENS

WHITE WATER LILY
CLOSES

SPOTTED CAT'S EAR OPENS 6 A M
EVENING PRIMROSE OPENS 6 P M

*The flower clock of Carolus Linnaeus depended on the opening
and closing of various flowers at regular times of the day. Linnaeus
did not include a sundial, but this interesting design permits the
observer to check the accuracy of the flowers relative to sun time;
it is within half an hour.*

That same year, in Geneva, Switzerland, the distinguished
and wealthy de Candolle family celebrated the arrival of a
son, who was christened Augustin Pyramus, and who was
destined to become one of the celebrated botanists of Europe.
De Candolle did his early studies in general natural science

in Geneva, but his deepest interest was in botany, and soon he went to Paris to work with Lamarck, Curier, and Geoffroy. In the following ten years his reputation grew and flourished, and he was elected a professor at Montpellier. Then, nearing forty, he returned to Geneva as professor of botany and member of the High Council.

He devoted much of his lifetime to the development of a new system for classifying plants which was a marked advance over that of Linnaeus, and he proposed a general scientific theory which in the logic of its design is striking even today. He conceived of nature as being controlled by four great forces, which he called attraction, affinity, life-force, and sensibility. And he held that these forces governed the phenomena observed, respectively, in physics, chemistry, physiology, and psychology. Among these broad fields, the physiology of plants held the most intense interest for de Candolle, and he devoted most of his study and investigation to the understanding of its problems. Toward the end of his career, he wrote a classic of its time, the three-volume *Physiologie Végétale*. Here we find his description of his own studies of the sleep movements of plants, which added much to the earlier insights gained by de Mairan and Duhamel.

Like his earlier compatriots, de Candolle began his investigations with a sensitive plant, but because of his expertise in plant classification it was second nature for him to report the material with which he worked by genus and species—*Mimosa pudica*. He confirmed the observations of de Mairan and Duhamel, and added proof that the times of opening and closing of the leaves were independent of temperature, and also independent of humidity (within the experimental accuracy then possible), and that they were clearly linked to the rising and setting of the sun. Then he took two truly imaginative steps.

First, he arranged a bank of six lamps so that they shone continuously on his plants with an intensity nearly equal to that of a sunless day. So far as these plants could see, night never came. Nevertheless, they continued to sleep at night and awaken in the morning in very nearly the same way as they had done in the normal cycle of day and night. De Candolle's careful observations revealed, however, one extremely important difference. *The clocks of the continuously lighted plants ran faster than normal!* They completed their daily cycles not in 24 hours, but rather in from 22 to 22½ hours. This kind of behavior, as we shall see later, is quite characteristic of living clocks when they are deprived of outside clues to the time of day.

Next, de Candolle asked himself what would happen if he scheduled his lamps so that the plants would be lighted at night and darkened during the day. They became "confused"—but only for a few days. And as the days passed, they began to fall into phase with the rhythm of the new lighting cycle and regularly closed their leaves at what seemed to them to be nightfall and opened them again when it seemed to be dawn. Soon they were running on a reverse light-dark regime as precisely as before.

Extending his observations to other kinds of plants, de Candolle found that different species of sensitive plants showed reactions similar to *Mimosa pudica*, although less markedly, while two species of oxalis refused to reverse their sleep-wake habits at all. De Candolle reasoned that different species of plants might need different brightnesses of light to change their reactions.[6] De Candolle's own publications contain no hint that he followed up this hunch, but in later chapters we shall see how accurately he guessed at some of the findings of modern botany.

THE LITTLE GREEN WORMS OF GEORGES BOHN

Were we now to follow the unfolding knowledge of the rhythms of plants in strictly chronological order, we should next join Charles Darwin in the work he did during his declining years on the power of movement in plants. However, Darwin was such a giant of science, and this latter work has

---

[6] Augustin Pyramus de Candolle: *Physiologie Végétale* (Paris: Béchet jeune; 1832), p. 859.

been so overshadowed by the impact of *The Origin of Species* and *The Descent of Man,* that we shall need a full chapter to place his work on the sleep of leaves in proper perspective. That chapter follows. Meanwhile we shall see how the movement of some tiny marine worms in a glass tube in Paris could accurately indicate the rise and fall of the tides in the Bay of the Seine.

In sharp contrast to such a man as Darwin, very many scientists spend their lives on a prosaic level, doing work that is useful, but hardly original enough to attract much notice. Nonetheless, their collective efforts have contributed notably to the advance of science. The point has been memorably expressed by two twentieth-century chemists, Gilbert N. Lewis and Merle Randall: "There are ancient cathedrals which, apart from their consecrated purpose, inspire solemnity and awe. Even the curious visitor speaks of serious things, with hushed voice, and as each whisper reverberates through the vaulted nave, the returning echo seems to bear a message of mystery. The labor of generations of architects and artisans has long been forgotten, the scaffolding erected for their toil has long since been removed, their mistakes have been erased, or have become hidden by the dust of centuries. Seeing only the perfection of the completed whole, we are impressed as by some superhuman agency. But sometimes we enter such an edifice that is still partly under construction; then the sound of hammers, the reek of tobacco, the trivial jests bandied from workman to workman, enable us to realize that these great structures are but the result of giving to ordinary human effort a direction and a purpose.

"Science has its cathedrals, built by the efforts of a few architects and of many workers."[7]

Yet even among these workers, a few will be found who make occasional brilliant contributions. The name of Georges Bohn, who worked in a Paris laboratory in 1903, is not to be

---

[7] G. N. Lewis and Merle Randall: *Thermodynamics and the Free Energy of Chemical Substances,* revised ed. (New York: McGraw-Hill Book Company; 1961), p. v. Used with permission of McGraw-Hill Book Company.

found in the biographical dictionaries of science. Nonetheless he made a fascinating discovery that anticipated some of the most striking work on rhythms of marine organisms by nearly half a century.

Leaving Paris in late July of that year, he spent more than a month on the shores of the Bay of the Seine, at St. Vaast-la-Hougue, between Cherbourg and Le Havre, and at St. Jacut-de-la-Mer. Let us explore a great university library to track down his report of this seashore adventure.

High on the sixth tier we find row on row of richly bound volumes containing the *Proceedings* of the French Academy of Sciences in Paris. The volumes, some three inches thick, are bound in grained brown leather stamped in gold. Taking down the one marked "1903" we see that the side coverings of the boards are of heavy black fabric, and the corners are reinforced with the same brown leather as the spine. On opening the book we are struck by paper of a quality that seems fine enough to have been handlaid, printed in a clean black type face of nineteenth-century French charm. We turn to page 756 and see the title, "*Sur les mouvements oscillatoires des Convoluta roscoffensis*," by Georges Bohn, submitted to the French Academy by M. Edmond Perrier. *Convoluta roscoffensis,* we realize, is a little marine flatworm that lives in the sand at the seashore and that owes its brilliant green coloring to the presence of symbiotic algae in its tissues. A free translation tells us something like this:

> These movements [that is, *les mouvements oscillatoires*] have been observed daily from the 25 July to the 30 September, at St. Vaast-la-Hougue, then at St. Jacut-de-la-Mer.
>
> The *Convoluta,* which are tubular and parasitized by green algae, live by the shore.[8] At each tide the water comes to cover them for a period of two and a half to five hours, depending on whether the tide is at the neap or the spring stage. They then move in the sand to different depths; when

---

[8] "Parasitized" is a literal translation; however, marine biologists today would say that the relationship is symbiotic, that is, the two organisms live together to their mutual benefit.

the tide goes out they come to form spots of intense green at the surface of the sand. The locations, sizes, and shapes of these spots change constantly.

The purpose of these movements is to avoid two dangers: washing away by the waves, and drying out.

Movement caused by the impact of the waves. *Convoluta* retreat toward the depths in following the vertical, and then come back up as soon as the shock stops after the retreat of the sea. During drying out, any agitation brings on a vertical descent immediately. If [in an experiment] one shakes a glass tube containing wet sand and some of the *Convoluta*, the latter descend, and then immediately come up again as soon as one stops the shaking. They form a green ring by which one can easily follow their movements. There are considerable differences in the sensibility to shock depending on the state of the tide. This is due to the fact that, apart from the movements provoked, there are those movements that correspond rigorously to those of the tide. If one places the tube in a quiet place, the green ring rises and descends alternately, occupying the highest position at the moment of the low tide and the lowest position at the time of the high tide. The synchronization has been seen to persist in an aquarium during fourteen consecutive tides. Such exterior influences as light, oxygen, and humidity, have not been able to change the rhythm. The movements also take place at night as well as during the day, and if one reverses the natural conditions of wetting and drying out, the movement is not reversed.

The movements take place within the sand, and then as soon as the animals have come to the surface, they continue to move along the sandy slopes. In a tidal basin the *Convoluta*, which have been under the sand when the tide went out—a little later during neap tides—gradually move up the slopes until the time of the low tide. Later, they leave in the same way. They therefore have the ability of following the movements of the tides at the seashore, even after they have been moved to an aquarium in Paris.

Movement caused by drying. The movements along the sandy shores are well seen on the beach. After the sea has retreated, the *Convoluta*, impelled toward the greatest height, move toward the shore and overrun the higher areas. This is later reversed and they descend the slope toward the sea and the damp depressions. But all of these shiftings can be influenced by the drying of the sand.

Now this drying is a function of the intensity of the sunlight. If it becomes more intense, the higher spots dry up, and the *Convoluta* move down the slope to reach damper spots. At the start of drying, the normal rise of the animals seems to be more difficult; toward the end the normal descent appears easier. In descending a sunny slope, *Convoluta* stop as soon as they cross into shade. The movement also stops if the light lessens suddenly during a descent. In some way, shade and light serve as warning signals, and the *Convoluta* always obey them. The same reactions take place whenever one removes the danger of drying by continued dampening. If the animals are placed in a cylinder full of water, they are seen to stop at the edges of all shadows and take the form of green lines.

This is not a question of phototropism. In reality there is no seeking of shadow or of light; no reversal takes place at the border between light and shade. They do not react in the same way during emergence from the sand; this is delayed by light.

Immediately following Bohn's paper, and in the larger type which the academy accorded him, M. Perrier added his own comments. They read in part as follows: "In summary, the *Convoluta* in the aquariums as well as in nature ascend and descend to the interior of the sand and along the sandy slopes. The movements are oscillatory, and the great voluntary movements that are synchronous with those of the tides overshadow the small movements brought on by the drying of the sand or by the changes in light intensity."

At this distance, we may well wonder why both Bohn and Perrier failed to stress the importance of the observation that the little green flatworms in an aquarium in Paris responded precisely to the rise and fall of the tides near the mouth of the Seine. Perhaps, in feeling the need to write an interpretative summary, Perrier sensed that Bohn had left some loose strings in his conclusions. In any case, the important point eluded them both.

WILHELM PFEFFER, FATHER OF PLANT PHYSIOLOGY

In sharp contrast to Georges Bohn stands the renowned German botanist Wilhelm Pfeffer. Pfeffer was a prodigious worker throughout his life, and produced a steady stream of scientific publications from the time he earned his Ph.D. in the University of Göttingen at the age of twenty until he died in 1920 at seventy-five. The new ideas that he put forward are so basic that scientists of today who work with living materials —plant, animal, or in the hazy world that lies between inorganic matter and living forms—depend on his teachings in the same way that they assume that the law of gravitation will hold tomorrow.[9]

Like de Candolle, Pfeffer observed that plants display daily sleep movements, their leaves being elevated by day and drooping at night. Pfeffer chose to work with young bean seedlings, a readily available and easily handled experimental material. Like de Candolle, Pfeffer found that the sleep rhythm persisted when light and temperature were kept constant, and his contribution in this area rests primarily on the fact that his demonstrations were more extensive than those of de Candolle, and his results are therefore more convincing.[1]

These observations, however, were trivial compared with two other contributions that he had made by 1880. One of these was a new concept, and the other was a new kind of apparatus. The concept was of the relationship between the osmotic pressures of solutions as a function of their concentrations—a relationship that is now known to explain the basic behavior of all living cells. The apparatus was the klinostat, which made possible for the first time the meaningful study of the effects of gravitational force on plant movements.

[9] *The Encyclopedia of Biological Sciences,* ed. Peter Gray (New York: Reinhold Publishing Corporation; 1961), p. 749.
[1] Frank A. Brown, Jr.: *Biological Clocks,* Biological Sciences Curriculum Study Pamphlet No. 2 (Boston: D. C. Heath and Company; 1962), p. 7.

We shall look at each of these contributions in some detail, but first, for perspective, let us make a rapid survey of Pfeffer's enormous range of accomplishments and interests.

Wilhelm Friedrich Philipp Pfeffer was born on March 9, 1845, in Grabenstein, near Cassel, Germany. After receiving his Ph.D. in botany he was appointed professor of botany at Bonn. In turn, he was professor of botany at Basel, Switzerland, Tübingen, Germany, and finally professor of botany and director of the Botanical Institute at Leipzig. He was elected a Fellow of the Royal Society of London, and a Member of the French Academy of Sciences. By the time he was thirty-five, he had completed and published the first edition of his famous *Handbuch der Pflantzenphysiologie.* A revised and enlarged second edition was published beginning in 1897, and the English translation of the second edition was published at Oxford in three volumes dated 1899, 1903, and 1905, under the title *Pfeffer's Physiology of Plants.* The scope of Pfeffer's interests is shown by the subjects with which he dealt in these volumes. Beginning with a statement of the general aims of physiology, he wrote of the structure and function of plant organs, the relation of plant cells to the structure of molecules, the mechanism of absorption and translocation of substances in plants, the movements of water both as fluid and as vapor, the food of plants, including carbon dioxide, organic foods, inorganic foods including especially nitrogen, "constructive" and "destructive" metabolism, and the functions of respiration and fermentation. Turning to the processes of growth and death, he considered the mechanics of growth, growth in relation to cell division, the elasticity and cohesion of the plant body, stresses and strains in tissues, the influence of external conditions on plant growth like light, temperature, magnetism and electricity, gravity and centrifugal force, chemical agents, and mechanical agencies. He then took up the causes of the specific shapes of plants and plant parts, questions of variation and heredity, periodicity of growth, the power of resistance to extremes, including causes of death. He further reviewed the broad field of plant movements, including the movements of plant parts that are

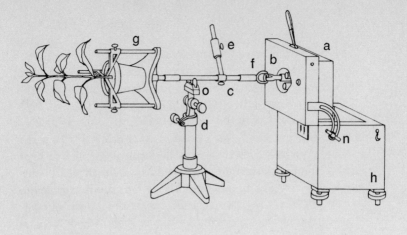

In Pfeffer's klinostat the movement is produced by a strong spring, the mechanism being attached to the lid (b) of the heavy box (h). One of the three axes on the upper surface of the lid is joined by the gimbal joint to the axis (c), which rotates on the friction wheels (o), and has a pot attached at (g). If the box is closed and a wood or cork plate attached to one of the axes, a pot may be placed on it and rotated around a vertical axis; or the lid may be inclined at various angles and fixed by the screw clamp (n) so that by means of the gimbal attachment the rod (c) is able to rotate around an oblique axis. The adjustable weight (e) is used to counterbalance any irregularities in the distributions of the materials being rotated. Pfeffer said that a simple form of klinostat could be made "from an ordinary American clock."

caused by gravity, sunlight, temperature, chemical agents, water, and electrical fields. He studied the locomotion of those forms of plant life that are capable of moving from place to place and found that they tend to seek out the most favorable conditions for nutrition and development. Finally, he reviewed the production of heat, light, and electricity by plants, and concluded with a short summary about the sources and transformations of energy in plants. In a word, he was thorough.

Out of this array of ideas, the two that are most closely related to modern studies of plant rhythms are Pfeffer's studies of osmosis in cells and the klinostat, with which he studied the effects of gravity on plant movements.

In pondering ways for investigating the effects of gravity, it occurred to Pfeffer that if he placed a potted plant in a horizontal position instead of upright and provided a mechanical drive to rotate the pot around its horizontal axis, the effects of gravity on the plant would be cancelled out. "If a seedling is slowly and steadily rotated on a klinostat," he wrote, "so that a revolution is performed in three to four minutes, the position of the plant is continually altered before any inductive stimulating action of gravity can be made manifest. For most plants, two to three revolutions per hour are sufficient, for at this rate practically no centrifugal force is exercised, while at the same time neither the shoot nor the root has time to make a curvature before its position is reversed. The heliotropic curvature (bending toward the sun) produced by unilateral illumination may also be prevented by rotation in a klinostat, so that both the heliotropic and geotropic action (bending due to gravity) may be eliminated if the plant is rotated about a horizontal axis at right angles to the direction of the illumination."[2]

Pfeffer rotated a plant of the scarlet runner bean in his klinostat and found that the absence of the gravitational field caused an alteration in the position of the leaves. However, he was vague about just what this alteration was, and it may be because he was dissatisfied with the outcome that he took the obvious step of turning the plant upside down. At once he saw that the leaves in the normal day position "passed in a few hours to a position resembling that assumed during the night."

A modern biologist would see so many questions Pfeffer might have put to his klinostat that today the results he actually obtained from it seem to lack substance. Nevertheless this same klinostat is the direct intellectual parent of a 1960 scientific expedition to the South Pole that was undertaken under the auspices of the U.S. National Science Foundation.

In contrast to this work with the klinostat, Pfeffer's con-

[2] Wilhelm F. P. Pfeffer: *Pfeffer's Physiology of Plants*, ed. and trans. Alfred J. Ewert, Vol. III (Oxford, 1905), p. 166.

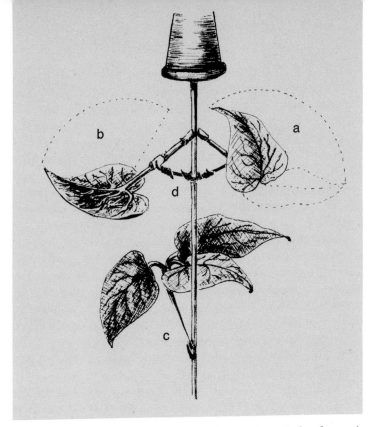

Phaseolus multiflorus *upside down. The petioles of the first pair of foliage leaves are fixed by the wire (d), so that only the pulvinus at the base of the lamina can curve. The leaf (a) is in the day position, while (b) is shown as it would be in the night position. The leaflets of the trifoliate leaf (c) rise to the normal light position by curvature of the basal pulvinus, and hence carry out the normal sleep movements.*

tribution to the understanding of the effects of osmotic pressure on the functions of living cells was truly experimental in a quite modern sense. A well-planned research program proceeds from the general to the particular. Thus a biologist interested, for example, in studying the little green worms of Georges Bohn would proceed from observations of the gross behavior of populations of these creatures moving up and down the sand, to a study of one individual, then to a single cell of an individual, and finally to the biochemical nature of the components of the cells themselves.[3]

Pfeffer laid the way for such programs through his own studies of the transfer of chemical substances through the plastic membranes that line cell walls. Speaking in general terms he wrote: "The character of the cell-wall and of the plastic membranes determines whether a given substance will penetrate to the interior of a cell, and any such substance will continue to be absorbed until a condition of equilibrium is reached, when all further absorption ceases. If, however, this condition of equilibrium is continually disturbed, continuity of absorption may result, and in this way, relatively large quantities of a particular substance may be absorbed from an extremely dilute solution, or of two given substances, one may be absorbed in large amount, the other scarcely at all. A continuous disturbance of equilibrium may be maintained, if the absorbed substances at once undergo a more or less marked chemical change or alteration into soluble or insoluble compounds of different character."[4]

Very well. But how does one study the action of such a cell membrane in the laboratory, especially with the limited kinds of apparatus that were known in 1880? For example, Pfeffer wanted to study the transfer of substances through cell walls in the roots of Indian corn. If one of these cells is magnified 375 times, it appears to be an inch high and half an inch wide. An object so small was impossible to manipulate by any technique that was then available; and Pfeffer was driven to devise something to simulate as nearly as possible the membrane of the cell. Specifically, he needed a membrane that would be permeable to small molecules like those of water, but that would not be permeable to larger molecules like those of sugar.

He was familiar with the famous Traube's cell (here "cell" does not mean the same thing as the cell of a plant

---

[3] Isaac Asimov's succinct summary states: "The cell theory is to biology about what atomic theory is to chemistry and physics." (*The Intelligent Man's Guide to Science* [New York: Basic Books, Inc.; 1960], p. 502.)
[4] Pfeffer: *Pfeffer's Physiology of Plants*, Vol. I, p. 89.

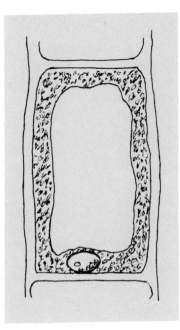

Within a single cell in a root tip of Indian corn (here shown magnified 375 times), the osmotic pressure forcing the cell membrane against the cell wall may rise to 150 pounds per square inch.

or animal), and he decided to prepare a few Traube's cells to see if they would serve his purpose. Pfeffer knew that Traube used precipitated copper ferrocyanide for his cells. This slimy brown precipitate feels like a thick solution of gelatin when one rubs it between the thumb and fingers. Pfeffer thought that it might act like a live cell membrane if it could be supported in the same way that a cell wall supports the membrane inside it.

For a mechanical support, then, Pfeffer tried unglazed porous clay cups, and "after the most scrupulous washing," he soaked them in solutions either of copper chloride or of copper acetate.[5] After thoroughly washing the outside of each cup with distilled water, he suspended them in a solution of potassium ferrocyanide. By this procedure he succeeded in forming an exceedingly thin and delicate membrane of the gelatinous copper ferrocyanide within the pores of the clay

[5] James Pierce Price: *Experimental Plant Physiology* (New York: Henry Holt and Company; 1931), p. 42.

cups. Finally he soaked the cups in many changes of distilled water, to remove all traces of unprecipitated soluble salts.

Soon he would know if these membranes would act like those of the living cells. At once he poured a dilute solution of sugar into one of his cups and dipped it into a beaker of distilled water. To his immense delight he saw that the level of liquid in the cup was slowly rising. Water was being drawn upward into the cup from below in its drive toward the solu-

Measuring the osmotic pressure of a sugar solution. Wilhelm Pfeffer used a simple apparatus like this to show that the osmotic pressure of the solution is proportional to the concentration of the sugar.

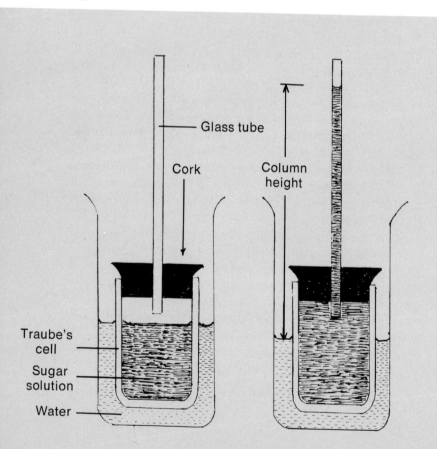

tion in the cup. But then was sugar passing downward through the membrane into the distilled water? Was it becoming sweet? It is hard to imagine that Pfeffer could have resisted dipping his finger into the water to taste it, but if he did so, he does not tell us. In any case, he followed his first trials with an extensive series of experiments that conclusively proved that the membrane permitted pure water to pass through it, but blocked the passage of sugar. Pfeffer had made a membrane that behaved like those of living cells, and by a technique so simple that he could prepare them for as many laboratory experiments as he wished.

As a scientist, Pfeffer's next concern was to see how he could measure the force that drove pure water through his membranes and into the sugar solution in the cup. The time had come to turn his qualitative observations into meaningful numbers.

For quantitative measurements, only the simplest of additions to his original cups was needed. He bored holes through corks that fitted the tops of the cups, and inserted long glass tubes into the holes, so that water being drawn into a cup could rise to any height needed to produce a downward pressure that would just balance the upward pressure of osmosis.

Pfeffer then planned a complete series of experiments, starting with dilute sugar solutions, and measuring the heights to which the solutions eventually rose in the glass tubes. Then, step by step, he increased the sugar concentrations of his starting solutions. He found that the more concentrated the sugar solution was, the higher the liquid column rose. If, then, he prepared a graph showing the concentration of sugar on one axis and the heights of solution in the glass tubes on the other axis, the points corresponding to the measurements fell on a straight line. This was conclusive proof that the osmotic pressure was proportional to the concentration. This is a quantitative relationship that scientists today take for granted as they go about their daily work. Today, too, cell physiologists can see a great deal farther into the inner workings of cells than Pfeffer could have dreamed of. But as Isaac Newton said: "If I have seen farther . . . it is by standing on the shoulders of giants."

THE COSMIC TIME-GIVERS OF SVANTE ARRHENIUS

In the late summer of 1884 Wilhelm Ostwald, a Latvian-born German professor of chemistry at the Polytechnic School at Riga, boarded a ship headed into the Baltic Sea for Stockholm. With him Ostwald carried a doctoral thesis that he had recently received from a young Swede named Svante Arrhenius.[6] Even though Ostwald had received the paper on a day that his wife had borne him a new daughter and he had had a toothache, he read it with mounting excitement at its brilliance.

But he felt that Arrhenius's contribution had been seriously misjudged. At the hands of the conservative faculty of the University of Uppsala the thesis had been graded "fourth class." Arrhenius had been granted his doctorate (to the accompaniment of the traditional cannon salute), but his work had been classed "non sine laude approbitur."

In Ostwald's judgment the ideas that Arrhenius had advanced, while certainly revolutionary, had such a ring of plausibility that they were quite likely to be right. If that proved true, they would surely open new fields in chemistry. Ostwald would not rest until he had confronted the laboratory director at Uppsala with his views.

Wilhelm Ostwald's mission lacked no credentials. His reputation was such that within two years he was to be

---

[6] This sketch relates only to Arrhenius as an iconoclast who in his later years tossed out an idea that was to develop into one of the stormiest controversies of modern biology. The general reader interested in battles of ideas will find a fascinating account of Arrhenius's career in Bernard Jaffe: *Crucibles, The Lives and Achievements of the Great Chemists* (New York: Simon and Schuster, Inc.; 1930), p. 219. More complete scientific detail and evaluation are to be found in Eduard Farber, ed.: *Great Chemists* (New York: Wiley-Interscience, Inc.; 1961), p. 1094; also "J. W.": "Svante Arrhenius, 1859–1927," *Proceedings of the Royal Society*, London, A, Vol. CXIX (1928), pp. ix–xix.

appointed professor of chemistry and director of the chemical laboratories at the University of Leipzig; later, he won the Nobel prize for chemistry.

After he left his ship at Stockholm Ostwald met young Arrhenius, and they talked of plans for winning acceptance of Arrhenius's ideas, or at the very least for gaining him a docentship at the University of Uppsala. Traveling north from Stockholm they agreed that it would be well for Arrhenius to stay clear of the first encounter with his former teacher.

Ostwald went alone into the laboratory of Per T. Clève, well aware that he would meet an older and far more conservative man, though a chemist of high attainment. Clève had won wide recognition for his work with the rare earths, had discovered the elements holmium and thulium, and had been appointed professor of chemistry and director of the laboratories at Uppsala. But he had little patience with new theories.

He had caused young Arrhenius one of his most painful moments as a graduate student, a moment that was especially painful because it followed so shortly after Arrhenius had had what he knew to be a flash of insight. "I got the idea on the night of the 17th of May in the year 1883," he wrote, "and I could not sleep that night until I had worked through the whole problem." Flushed with excitement he burst into Clève's laboratory to set forth his theory. Clève stared at him out of skeptical old eyes. "You have a new theory? That is very interesting. Goodbye."

The new theory that Clève so casually dismissed was later to become universally hailed as the electrolytic theory of dissociation; it was to receive acceptance by chemists everywhere; and it was to earn for Arrhenius the Nobel prize for chemistry in 1903.

In simple terms, the theory of electrolytic dissociation holds that when inorganic substances are dissolved in water, the molecules of these substances leave their original form and break apart into separate electrically charged particles, called ions. When chemical reactions take place in solutions, the reactions are between these ions only.

Svante Arrhenius, winner of the 1903 Nobel prize in chemistry. Before the turn of the century he had suggested that biological rhythms might be caused by tides in the cosmic forces that surround the earth.

If, for example, we take a spoonful of common salt and dissolve it in water, the following occurs:

$$NaCl + H_2O \longrightarrow Na^+ + Cl^- + H_2O$$

white crystals	liquid		colorless particle	colorless particle	liquid

The positive charge of the sodium and the negative charge of the chlorine are responsible for a dramatic difference from the properties of the elements without the charges. Elemental sodium is a light, white metal, so soft that it can be cut with a dull knife, and so reactive with water that it cannot be stored in ordinary air that contains even traces of water vapor. Elemental chlorine is the greenish-yellow poisonous gas that was used with such devastating effect in World War I. But when the atoms of these two elements are charged electrically and they occur together in aqueous solution, we have nothing but ordinary salt water. However, the sodium ions and the chlorine ions are ready to react on the instant with other ions that may be added.

What more understandable, then, that a conservative gentleman of the old school like Clève should look upon Arrhenius as a young upstart; should feel that it would be an irresponsible act to appoint him to a docentship at Uppsala; and should suspect that the new theory bordered on insanity.

Ostwald's passionate arguments—clear and logical though they were—failed to change Clève's mind, and later Ostwald wrote: "I can still plainly recall the scene in the chemical laboratory in Uppsala, where the head, himself an eminent chemist, heatedly asked me, pointing to a beaker containing an aqueous solution—'And you ask me to believe that sodium atoms are swimming around there in this fashion?' When I agreed he quickly looked at me in such a manner as if he had considerable doubt about my chemical rationality." But in the end Clève relaxed his position somewhat, and Ostwald was able to conclude: "However, this did not prevent the esteemed older colleague from receiving us both with true Swedish hospitality, and later he did what he could to remove the difficulties surrounding the granting of the docentship."

Thus the career of a scientist who dared to challenge the fixed ideas of the Establishment was launched in a storm, and it met heavy weather for years afterward. Nonetheless Arrhenius fought on, aided by few chemists other than Ostwald himself, and the Dutch professor Jacobus van't Hoff. And at very long last he won. In 1903 he was awarded the Nobel prize for chemistry, and soon afterward the Berlin Academy of Sciences sought him as professor of chemistry. But to better that offer, King Oscar II of Sweden founded the Nobel Institute for Physical Chemistry at Stockholm, and saw to it that the Swedish Academy appointed Arrhenius its director. ". . . at Experimentalfältet, in a pretty park in the neighborhood of the town, a small laboratory was erected, with an official residence attached. The laboratory was inaugurated in 1909. Here, with an assistant, and a few research workers as guests, Arrhenius could work and write under ideal conditions on such problems of physical chemistry, physiological chemistry, immunochemistry, meteorology, and cosmic physics as might please him."[7]

One of the problems that had pleased him a bit earlier was the question of how cosmic influences might affect the physiological behavior of plants, animals, and men. He attacked this problem with his usual daring, taking a few tables of data, studying them intently for a while, and then flashing forth a new theory to explain them.

Some of the tables he studied gave data on the periodic occurrence of bronchitis, and on the periodic variations in birth and death rates. He noted the approximately lunar periodicity of menstruation in the human female and thought that his numbers pointed to a period length of 25.929 days.[8] He also had tables that suggested a periodic onset of epilepsy. He compared graphs showing high and low frequencies of these physiological changes with graphs that showed changes in the electrical charges in the atmosphere. Finally, he thought that he could see a definitive relationship between the two. Such a correlation would suggest that one of the phenomena might be the cause of the other, and Arrhenius himself concluded that variations in atmospheric electricity *caused* variations in the menstrual cycle, in the occurrence of bronchitis and epilepsy, and even in birth and death rates in various parts of the world.

More, he produced a biochemical explanation for it.

---

[7] "J. W.": "Svante Arrhenius," p. xviii.
[8] A modern statistician would ask Arrhenius what the confidence limits of such an apparently exact number might be.

"The physiological influence of atmospheric electricity," he wrote, "which has been known for a long time in plants, may have a broad influence on the whole of living nature. A high electrical tension in the air may cause chemical reactions which have an effect on the organisms. This would affect the reproductive cycle of palolo worms and of other animals of different characteristics. And it is clear that atmospheric electricity has a marked influence on nervous disorders."[9]

With that, Arrhenius went on to other problems that held greater interest for him at the time. He had offered his own suggestion, and he left it to others of a more methodical nature to fill in the details.

But no one with that turn of mind came forward. Nothing of consequence was done on the effect of cosmic forces on the periodicities of living organisms for half a century. And when the problem was finally directly attacked, it was by no prosaic mind. It was by the mind of a man quite as daring, quite as willing to question the theories of the Establishment, and quite as unwavering in his fight for his own views, as was Arrhenius. Further, unlike Arrhenius, that man had the criticisms of modern statisticians and the skepticism of modern scientists to deal with, for he did not begin his work until some 50 years later. That man is Frank A. Brown, Jr., Morrison Professor of Biology at Northwestern University, whose controversial theories about the influence on living clocks of the subtle mysterious geophysical forces that surround the earth fired one of the sharpest controversies in modern biology. In his rightful historical place we shall follow Brown in his long and spirited struggle with a formidable array of his peers.

First, however, let us see how Charles Darwin's intensive work on plant growth has influenced all later studies of rhythms.

---

[9] Svante Arrhenius: *"Die Einwirkung kosmischer Einflüsse auf physiologische Verhältnisse," Skandinavisches Archiv für Physiologie,* Vol. VIII (1898), pp. 367–416 and 5 plates.

*"As for myself,
I believe that I have
acted rightly in steadily
following and devoting
my life to science."*

CHARLES DARWIN,
1879

4. Darwin: Why Plants Sleep

A few scientists in history have given the world ideas so revolutionary that after them man's concept of nature could never be the same. The greatest of these giants were Galileo, Newton, Einstein—and Charles Darwin. Darwin's work towers in two respects: it was so broad that its implications still influence all fields of human endeavor; and it was so seminal that his original works remain a mine of ideas for fresh study.

In the words of Sir Gavin de Beer, ". . . professional biologists and geologists would be astonished to find that in addition to the demonstrations of fact and theory for which [the works] are justly famous, they contain an inexhaustable supply of problems for research of central importance at the present day."[1]

[1] Sir Gavin de Beer: *Charles Darwin, Evolution by Natural Selection* (New York: Doubleday & Company, Inc.; 1964), p. vi.

Toward the end of his life Darwin turned again to his work on plant growth, and in his autobiography he says that he took more pleasure in this than in almost any other phase of his work. He gave special attention to de Mairan's phenomenon of the daily movements, up and down, of the leaves of plants, and he carried the work of Duhamel and de Candolle to new levels of experimental sophistication. We shall look at these experiments shortly.

But they are best seen in relation to the central theme of Darwin's work, and we shall first briefly trace his abiding concern with natural selection.

On November 24, 1859, John Murray, Publisher, of 50 Albemarle Street, London, placed on sale a five-hundred-page "abstract": *On the Origin of Species by Natural Selection, or the Preservation of Favoured Races in the Struggle for Life,* by Charles Darwin. The twelve hundred and fifty copies of the first edition were sold that day, and Murray immediately ordered three thousand more.

So began the history of the "most important publication of the nineteenth century."[2]

So also began the bitterest scientific-ethical-religious controversy of the nineteenth century—and beyond. Indeed, it was twentieth-century society that carried it to the point that the state of Tennessee saw fit to prosecute John Thomas Scopes for violating its statute holding it to be "unlawful for any teacher in any of the universities, normals, and other public schools of the state, to teach any theory that denies the divine creation of man as taught in the Bible, and to teach instead that man has descended from a lower order of animals." In July, 1925, Scopes was convicted, William Jennings Bryan suffered such withering cross-examination at the hands of the brilliant Clarence Darrow that his death five days later may have been caused by it, and the press of the day spewed

[2] G. H. Parker, quoted in Jacques Barzun: *Darwin, Marx, Wagner: Critique of a Heritage* (Boston: Little, Brown and Company; 1941), p. 94.

forth two million words of salacious copy for the world to wallow in. Of these two million words, not one dealt with whether the theory of natural selection was right or wrong. The judge had barred all testimony on this from the trial.

Of such a development, Darwin never dreamed. He had worked for twenty years on painstaking observations and experiments, studied hundreds of publications by others, and collected and classified voluminous folios of notes on geology, botany, zoology, and what we know today as ecology. When he came to the actual writing of the "abstract" it cost him "thirteen months and ten days' hard labour."[3] As a fifth edition neared publication in 1869, he wrote to J. D. Hooker, world authority on botany and long-time friend: "If I lived twenty more years and was able to work, I should have to modify the *Origin,* and how much the views on all points would have to be modified!"[4] What a far cry from the utterances of bigots—then and later.

More than ten years after Murray published the first twelve hundred and fifty copies of the *Origin of Species,* Thomas Henry Huxley, biologist, lecturer, and member of ten royal commissions wrote: "The gradual lapse of time has now separated us by more than a decade from the date of the publication of the *Origin of Species;* and whatever may be thought or said about Mr. Darwin's doctrines, or the manner in which he has propounded them, this much is certain, that in a dozen years the *Origin of Species* has worked as complete a revolution in Biological Science as the *Principia* did in astronomy, [and it has done so] because in the words of Helmholtz, it contains 'an essentially new creative thought.' And, as time has slipped by, a happy change has come over Mr. Darwin's critics. The mixture of ignorance and insolence which at first characterized a large proportion of the attacks

[3] Charles Darwin: *The Autobiography of Charles Darwin, and Selected Letters,* ed. Francis Darwin (New York: Dover Publications, Inc.; 1958), p. 44.
[4] Charles Darwin: *The Autobiography of Charles Darwin, 1809–1882, With Original Omissions Restored,* ed. Nora Barlow (New York: W. W. Norton & Company, Inc.; 1969), p. 3.

with which he was assailed has no longer the sad distinction of anti-Darwinian criticism."[5]

Next to the *Origin of Species* stands Darwin's *The Descent of Man, and Selection in Relation to Sex,*[6] which was first published in 1871. The writing of the *Descent* took three years, and in another of Darwin's letters to Hooker, he said: "I finished the last proofs of my book a few days ago; the work half-killed me, and I have not the most remote idea whether the book is worth publishing."[7] History was a better judge of the value of the work than its author, for the two books together represent a monument to man's capacity for inductive reasoning.

In the *Descent* Darwin developed further some of the views that he had set forth in the *Origin,* and he went on to explore what should logically follow if the principles of natural selection were applied to human beings. He concluded —even though it would be "highly distasteful to many"—that man's ancestors were probably monkey-like creatures related to the ancestors of contemporary orangutans, chimpanzees, and gorillas. And he supported the thesis with example added to example.

An especially clear modern summary of the five main points that Darwin made in the two works is available to college students today: "*First,* all existing vegetable and animal species are descended from earlier and, generally speaking, more rudimentary forms. *Second,* the variation in species has come about because the environment and the use or disuse of organs have brought about changes in structure that are inherited. *Third,* in the struggle for survival, the fittest win out at the expense of their rivals because they succeed in adapting themselves best to their environment. *Fourth,* differentiation among the species is also brought about by

[5] Charles Darwin: *Autobiography,* ed. Francis Darwin, p. 287.
[6] In his study of Darwin's work on selection through sexual preference, Jacques Barzun (*Darwin, Marx, Wagner,* p. 68) could not resist quoting Weismann: "It arises from the rivalry of one sex, usually the male, for the possession of the other, usually the female."
[7] Charles Darwin: *Autobiography,* ed. Francis Darwin, p. 286.

sexual selection, which Darwin declared is 'the most powerful means of changing the races of man.' *Finally, some variations seem to arise spontaneously, a view of Darwin's which pointed toward the doctrine of mutation."*[8]

The revolutionary nature of Darwin's work, and especially the overly emotional reaction to it displayed by the press and the public, have overshadowed Darwin's own development of the basic strategy of science. Let us look into this trend more closely, for it well illustrates a parallel trend that was taking place in all of science in Darwin's time, and that has continued at an accelerated rate ever since.

Darwin's productive years spanned most of the Victorian age.

Two days after Christmas of the year 1831, this well-born young Englishman, full of mutton chops and champagne, boarded H.M.S. *Beagle* to serve as its naturalist during a world cruise that was to last five years. Fifty years later, as a Corresponding Member of the Botanical Section of the French Institute, he saw *The Power of Movement in Plants*, his last important book, placed on sale by John Murray. The fifty-year epoch may be divided very roughly into three major periods: a period of collection, a period of synthesis and analysis, and a period of specialization.

During the voyage of the *Beagle*, Darwin collected specimens and observations about a bewildering variety of plants and animals in disordered array.

During the period of thinking and writing from which the *Origin* and the *Descent* emerged, he organized this welter of material into a unified general theory—a landmark of inductive reasoning.

In his later years, he sought to test the soundness of the general theory by limiting his experimental materials to plants.

---

[8] T. Walter Wallbank, Alastair M. Tayler, and George Barr Carson, Jr.: *Civilization Past and Present*, 5th edn. (Glenview, Ill.: Scott, Foresman and Company; 1965), Vol. II, p. 326.

While he never approached the specialization of a modern plant physiologist, who might spend years studying the behavior of a single species, he nevertheless moved generally away from the descriptive, qualitative approach, and toward the quantitative analytical approach. And characteristically he contributed new concepts that were to lead others—decades later—to deeper understandings of why living things behave as they do.[9]

First, how did Darwin's work with plants affect general plant physiology? For answer let us turn to a modern botanist, Dr. Barbara Gillespie Pickard, of the Biology Department and the Research Laboratory of Electronics at the Massachusetts Institute of Technology. In her preface to the 1966 republication of Darwin's *Power of Movement in Plants* she said: "If asked to choose the scientific report which most clearly marks the beginning of the modern study of plant growth, a great many botanists would select Charles Darwin's *The Power of Movement in Plants.* Darwin's interest in the growth and turgor movements of higher plants [that is, movement of the plant parts caused by variations of osmotic pressure within the plant cells, as discussed in Chapter 3] was broad indeed, as one might expect of the celebrated evolutionist. His book covers the sleep movement of leaves which help prevent excessive radiation loss at night, the oscillations of vines which bring them into contact with support, and arching of the hypocotyl which strengthens it for its push through the soil, the spreading of leaves to catch the sunlight, the bending of flower stalks and the burying of certain seed capsules, random mutations, the evasion of stones and other obstacles by roots, and the influence of light, gravity, and

---

[9] Because Darwin's research on plant movements includes two major forward steps concerning biological rhythms, this chapter emphasizes that work. For an account of Darwin's life as a whole see Ruth Moore: *Charles Darwin: A Great Life in Brief* (New York: Alfred A. Knopf, Inc.; 1955).

moisture in orienting plants in optimal positions for further growth."[1]

As early as 1863 Darwin had written to Professor Asa Gray of Harvard about the fascination that plants held for him: "My present hobby-horse I owe to you, viz. the tendrils: their irritability is beautiful, as beautiful in all its modifications as anything in Orchids. About the *spontaneous* movement (independent of touch) of the tendrils and upper internodes, I am rather taken aback by your saying, 'is it not well known?' I can find nothing in any book which I have. . . . The spontaneous movement of the tendrils is independent of the movement of the upper internodes, but both work harmoniously together in sweeping a circle in one fixed direction. It is surprising to watch the Apocyneae with shoots 18 inches long (beyond the supporting stick) steadily searching for something to climb up. When the shoot meets the stick, the motion at that point is arrested, but in the upper part is continued; so that the climbing of all plants yet examined is the simple result of the spontaneous circulatory movement of the upper internodes. Pray tell me whether anything has been published on this subject? I hate publishing what is old; but I shall hardly regret my work if it is old, as it has much amused me. . . ."[2]

In the summer of 1877 Darwin at sixty-eight was so ill that on many days he did not leave his sofa. But lying there he could watch the slowly rotating movement of a plant tendril as it grew and groped for support. His mind was as clear and active as ever, and as he watched his plants he began to formulate an ambitious program designed to explore the relationship between the various movements of plants and

---

[1] Charles Darwin and Francis Darwin: *The Power of Movement in Plants*, with a "Preface" by Barbara Gillespie Pickard, unabridged republication of the 1881 edition (New York: Plenum Publishing Corporation: Da Capo Press; 1966), p. v.
[2] Charles Darwin: *Autobiography*, ed. Francis Darwin, p. 333.

their fitness for survival. By fall he knew the course he would take. He wrote to friends all over the world and gathered a collection of more than three hundred plant species. Those that could not stand the cold of an English winter he placed in the greenhouse that he had had built years earlier. His son Francis assured him of his full-time assistance. And in a letter to W. Thiselton Dyer, Darwin exclaimed: "I am all on fire at the work." It was work that was to last for three years.

The full report of the work of those years, as Darwin gave it in *The Power of Movement in Plants,* required nearly six hundred pages of meticulous detail about the aims, methods, and results of literally thousands of experiments. Some of them failed, some gave inconclusive results, and many were truly beautiful. Our interest centers on two of the latter which had the greatest influence on the research of the scientists who followed.

First we shall see how Charles and his son Francis learned that light falling from one side determines the direction in which canary grass seedlings bend—work that later led Went and Cholodny to the first understanding of plant hormones. Second we shall see how Darwin used Virginia tobacco to help explain why plant leaves sleep.

The Darwins directed their experiments with canary grass seedlings to the effects of light on small localized areas, and how these effects are transmitted to other parts of the seedling. "Whilst observing the accuracy with which the coleoptiles[3] of this plant become bent towards the light of a small lamp," they wrote, "we were impressed with the idea that the uppermost part determined the direction of the curvature of the lower part. When the coleoptiles are exposed to a lateral light, the upper part bends first, and afterwards the bending gradually extends down the base, and, as we shall presently see, even a little beneath the ground."

[3] In the original text, Darwin wrote "cotyledon" rather than "coleoptile," since that was the accepted term in his day. Modern usage prefers "coleoptile" for the first leaf to appear above ground in the seedlings of grasses; we have therefore taken the liberty to use "coleoptile" throughout.

Given the hypothesis that the light-sensitive area in a growing shoot is localized somewhere near its tip, how would one test the idea? Obviously by covering the tip in some way, so that light could not reach it. This is just what Darwin did, in several ways: "To test our belief that the upper part of the coleoptiles of *Phalaris* [canary grass], when exposed to lateral light, regulates the bending of the lower part, many experiments were tried; but many of our first attempts proved useless from various causes not worth specifying." Darwin tried using caps of very thin tin foil, painting the tips directly with either India ink or with varnish containing black pigment, using very thin glass tubes painted black on the inside, or even amputating part of the tip. He concluded that the glass tubes gave the most reliable results, because parallel experiments could be done with clear glass tubes to cancel out any effects that the slight weight of the tubes might have— the controlled experiment again:

> . . . caps were made of tubes of the thinnest glass, which when painted black served well, with the one great disadvantage that the lower ends could not be closed. But tubes were used which fitted the coleoptiles almost closely; and black paper was placed on the soil round each, to check the upward reflection of light from the soil. Such tubes were in one respect far better than caps of tin foil, as it was possible to cover at the same time some coleoptiles with transparent and others with opaque tubes; and thus our experiments could be controlled.

> We will begin with the glass tubes. The summits of nine coleoptiles, differing somewhat in height were enclosed for rather less than half their lengths in uncolored or transparent tubes; and these were then exposed before a southwest window on a bright day for eight hours. All of them became strongly curved toward the light, in the same degree as many other free seedlings in the same pots; so that the glass tubes certainly did not prevent the coleoptiles from bending toward the light. Nineteen other coleoptiles were, at the same time, similarly enclosed in tubes thickly painted with India ink. On five of them, the paint, to our surprise, contracted after exposure to the sunlight, and very narrow cracks were formed, through

which a little light entered; and these five cases were rejected. Of the remaining fourteen coleoptiles, the lower halves of which had been fully exposed to the light for the whole time, seven continued quite straight and upright; one was considerably bowed to the light, and six were slightly bowed, but with the exposed bases of most of them almost or quite straight. It is possible that some light may have been reflected upwards from the soil and entered the bases of these seven tubes, as the sun shone brightly, though bits of blackened paper had been placed on the soil round them. Nevertheless, the seven coleoptiles which were slightly bowed, together with the seven upright ones, presented a most remarkable contrast with the many other seedlings in the same pots to which nothing had been done. The blackened tubes were then removed from ten of these seedlings, and they were now exposed before a lamp for eight hours; nine of them became greatly, and one moderately, curved toward the light, proving that the previous absence of any curvature in the basal part, or the presence of only a slight degree of curvature there, was due to the exclusion of light from the upper part.

After describing in considerable detail a similar series of experiments in which they used tin foil to cap the tips of the canary grass shoots, the Darwins summarized their conclusions: "From these several sets of experiments, including those with the glass tubes, and those when the tips were cut off, we may infer that the exclusion of light from the upper part of the coleoptiles of *Phalaris* prevents the lower part, though fully exposed to a lateral light, from becoming curved. . . . We must therefore conclude that when seedlings are freely exposed to a lateral light *some influence is transmitted from the upper to the lower part, causing the latter to bend.*"[4]

Now it is nearly a century since Darwin drew his conclusion that "some influence" transmitted from the upper to the lower part of a seedling shoot caused it to bend toward the light. Darwin took great pleasure in observations that might turn out to have wide implications, and Dr. Pickard

[4] The italics are mine. The "influence" was later shown by Went and Cholodny to be the plant hormone auxin, but only after another fifty years had passed.

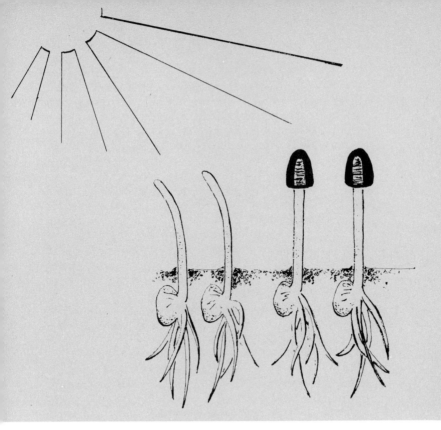

Plants bend toward a light source coming from a single direction. By covering the upper tips of canary grass seedlings with black paper, blackened glass tubes, or tin foil, Charles and Francis Darwin showed that the "influence" that causes the bending originates in the tip.

is no doubt right when she says that "he would have been deeply gratified to learn how general have been the applications of the ideas developed from the work on [the coleoptile]. Auxin [a plant hormone] is now known to control or influence almost every phase of higher plant growth and development."[5]

In the springtime auxin moves down a tree trunk and stimulates division of the cells that cause the trunk to grow outward. Woody cells develop different forms in response to

[5] Barbara Gillespie Pickard in Charles Darwin and Francis Darwin: *The Power of Movement in Plants*, p. xii.

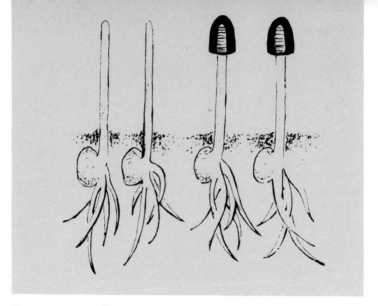

Canary grass seedlings exposed to light coming uniformly from all directions (as indoors or in a greenhouse) do not bend, whether their tips are shielded from the light or not.

the functions they must perform, and higher levels of auxin on the lower sides of bent trees cause the cells to develop especially strong walls. High auxin levels lead to the formation of the tiny valvelike pores on the surfaces of leaves that regulate the escape of water vapor and the absorption of carbon dioxide from the air. Roots produce their primary hairs under the influence of auxin. Buds are held dormant by it. The fall of leaves is inhibited by it. Many stages of fruit development are under its control, and the enlargement of fruits depends on it. Finally, auxin controls a very wide range of plant movements. Thus it will be apparent that while Darwin did not devote his primary attention to the rhythmic responses of canary grass to light intensity or directional cycles, his experiments had very far-reaching implications for the future.

At the same time it will also be clear that various external stimuli are responsible for various plant movements. Botanists divide these movements into two kinds: nastic movements, and tropisms. A nastic movement has no relation to the direction from which a stimulus comes, and the direction in which the organ moves is determined either by the structure of the organ, or by its relation to the plant as a whole. A tropism is a bending that is determined by the direction from which the stimulus comes, but the bending may be either toward the stimulus or away from it. Thus the primary shoots of Darwin's canary grass bent toward the light source, but when the light was removed, they returned to the vertical under the influence of apogeotropism (a bending away from the attraction of the earth).

We have seen that Darwin's main interest in studying plant movements was to discover an evolutionary relationship among them. He could not identify it in his canary grass studies because at least a half century's progress in plant physiology would be needed before the underlying mechanism could become clear. In contrast, the evolutionary relationship associated with the sleep of leaves seemed manifest, and Darwin found the sleep of leaves endlessly fascinating, both to watch and to theorize about.

His experimental methods were elegant in their simplicity, particularly in contrast to today's lavish investment in recording photocells, computerized data analyzers, and output devices that automatically draw curves of experimental parametric values in the appearance of three dimensions. Listen to Darwin's own description of what he was able to do with a bit of glass fiber, a white card, a wooden stick, a dot of sealing wax, a sheet of glass, and a bottle of India ink:

Methods of Observation. The movements, sometimes very small and sometimes considerable in extent, of the various organs observed by us, were traced in the manner which after many trials we found to be the best, and which must be described. Plants growing in pots were protected wholly from the light, or had light admitted from above, or on one side as the case might require, and were covered above by a large horizontal sheet of glass, and with another vertical sheet on one side. A glass filament, not thicker than a horsehair, and from a quarter to three quarters of an inch in length, was affixed to the part to be observed by means of shellac dissolved in alcohol. The solution was allowed to evaporate, until it became so thick that it set hard in two or three seconds, and it never injured the tissues, even the tips of tender radicles, to which it was applied. To the end of the glass filament an excessively minute bead of black sealing wax was cemented, below or behind which a bit of card with a black dot was fixed to a stick driven into the ground. The weight of the filament was so slight that even small leaves were not perceptibly pressed down. . . . The bead and the dot on the card were viewed through the horizontal or vertical glass plate (according to the position of the object), and when one exactly covered the other, a dot was

Darwin's elegantly simple method of recording leaf movement. With his eye to the right of the vertical glass sheet, he placed the black dot on the white card at (a) and the dot of sealing wax on the end of the glass filament at (b) (which had been cemented to the leaf tip) in exact alignment; he then made an India ink mark on the outside of the glass sheet at (c). As the leaf moved, he repeated the operation, making new dots on the glass, say at (c_i), (c_{ii}), (c_{iii}), etc. Finally, he copied the dots on tracing paper and joined them with ruled lines with arrows showing the direction of movement.

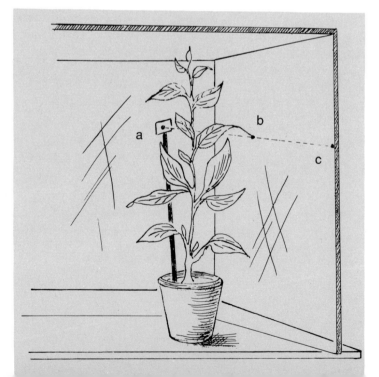

made on the glass plate with a sharply-pointed stick dipped in thick Indian ink. Other dots were made at short intervals of time and these were afterwards joined by straight lines. The figures thus traced were therefore angular; but if dots had been made every one or two minutes, the lines would have been more curvilinear, as occurred when radicles were allowed to trace their own courses on smoked glass plates. To make the dots accurately was the sole difficulty, and required some practice. Nor could this be done quite accurately when the movement was much magnified, such as thirty times and upwards; yet even in this case the general course may be trusted. To test the accuracy of the above method of observation, a filament was fixed to an inanimate object which was made to slide along a straight edge and dots were repeatedly made on a glass plate; when these were joined, the results ought to have been a perfectly straight line, and the line was very nearly straight.[6] It may be added that when the dot on the card was placed half an inch below or behind the bead of sealing wax, and when the glass plate (supposing it to have been properly curved) stood at a distance of seven inches in front (a common distance), then the tracing represented the movement of the bead magnified fifteen times.

After dots had been made on the glass plates . . . they were copied on tracing paper and joined by ruled lines, with arrows showing the direction of the movement. The nocturnal courses are represented by straight broken lines. The first dot is always made larger than the others, so as to catch the eye, as may be seen in the diagrams.[7]

Using this method Darwin traced the movements of the leaves of a variety of plants under many different conditions. He used strong light and dim light, diffuse light and directional light, and finally no light at all (except for brief observation periods) both at night and in artificial darkness during the day.

---

[6] In principle this arrangement was somewhat similar to the smoked-glass plates on which Darwin induced growing root tips to trace their own courses—courses that always seemed to him "wonderful" and that he "took an especial delight in showing."
[7] Charles Darwin and Francis Darwin: *The Power of Movement in Plants,* p. 6.

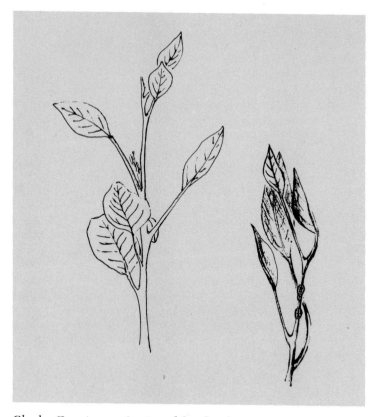

Charles Darwin was fascinated by the sleep movements of leaves, and had drawings made from photographs taken of tobacco leaves expanded during the day, left, and asleep at night, right.

An experiment with a plant of Virginia tobacco was typical of the records of leaf movement that he made. In the book, he first showed a composite drawing made from two photographs of Solaneae tobacco, one with the leaves spread out during the day and the other with the leaves asleep at night. Along with it, he showed a reproduction of a tracing he had made of the movement—during a period of sixty-five hours—of the fifth leaf above the coleoptile of a Virginia tobacco plant. The major movement of the leaf was in the shape of a large ellipse, reaching its lower point between 3:00 and 5:00 P.M., and its higher point between 10:00 and

11:00 A.M. On one day, the leaf did not extend as fully as on two other days, and then it zigzagged a little. Darwin was impressed with the simplicity of movement of this particular leaf, since it was quite different from the very complicated movements that other plants had shown. "The tracing is remarkable . . . from its simplicity and the straightness of its lines."

Most of the tracings from other plant varieties (he listed 55 varieties whose coleoptiles sleep and 86 varieties whose

Movement of a single leaf of Virginia tobacco as Darwin recorded it from 3:00 P.M. on July 10th to 8:10 A.M. on July 13th. The starting point was the larger dot at lower left, and the arrows show the direction of movement. Solid lines represent daytime movement, and dotted lines, movement at night. The original trace was about 1.5 times that shown.

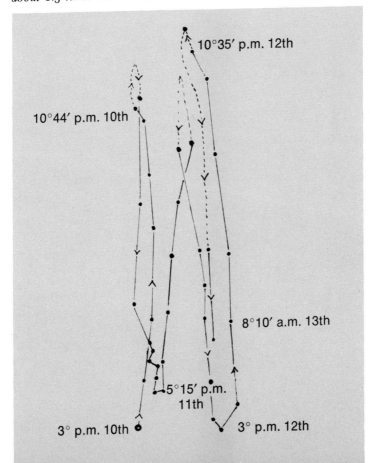

leaves sleep) showed much more complicated sleep movements. Sometimes the leaves rose and fell by tiny jerks, sometimes by longer movements like waves. Often the rise and fall was accompanied by a twisting motion. Sometimes the leaves slept by rising up and folding together, and sometimes by sinking down. Sometimes the outer leaves curled over the tenderer inner leaves as a flower closes.

All this led Darwin to make a characteristic generalization. No matter what position the leaves assumed when fully asleep, *the attitude was such as to expose the smallest possible leaf surface to the night sky.* Here was the clue to the evolutionary relationship that he sought!

Through natural selection, the plants that he was now observing possessed an ideal mechanism for protecting themselves from cold: ". . . there can be no doubt that the position of the leaves at night affects their temperature through radiation to such a degree that when exposed to a clear sky during a frost, it is a question of life and death."

This was the hypothesis. The next step was to test it. From his greenhouse Darwin took plants of oxalis, peanut, cassia, sweet clover of six varieties, a lotus, and a pepperwort, and exposed them on frosty March nights to temperatures of five or ten degrees below freezing, for periods of half an hour up to five hours. He pinned test leaves out in a horizontal position with fine insect-mounting pins secured to cork sheets beneath; the upper surfaces of the leaves were thus fully exposed to the night sky. As controls, he left most of the leaves in the pots in their normal sleeping positions.

The results all pointed to the same general conclusion, though they did show substantial variations. The temperatures were different on different nights; the sky was sometimes clear and sometimes hazy; the humidity of the air varied from night to night, and different amounts of dew collected on the leaves. Some of the plants were so sensitive, or the nights were so cold, that most of the leaves were killed, both the experimentals and the controls. Some plants were so hardy that none of the leaves were affected. But for the most part the leaves that were pinned out horizontally were severely damaged or killed, and those in the normal sleeping position survived.

While a statistical analysis would have supported Darwin's conclusion better than his somewhat qualitative generalizations, no one now doubts that "leaves exposed to a clear sky, with their blades compelled to remain horizontal, suffered much more from the cold than others that were allowed to assume their proper vertical position." Or doubts that another mechanism had been demonstrated through which the fittest might survive.

Except for a "little book" on how earthworms build organic mold in the soil, this was the final work of Darwin's life. On May 1, 1881, he summed up his thoughts about the longer book in these words:

"In 1880 I published, with Frank's assistance our *Power of Movement in Plants.* This was a tough piece of work. The book bears somewhat the same relation to my little book on *Climbing Plants* which *Cross-Fertilization* did to the *Fertilization of Orchids;* for in accordance with the principle of evolution it was impossible to account for climbing plants having been developed in so many widely different groups unless all kinds of plants possess some slight power of movement of an analogous kind. This I proved to be the case; and I was further led to a rather wide generalization, viz., that the great and important class of movements, excited by light, the attraction of gravity, &c., are all modified forms of the fundamental movement of circumnutation. It has always pleased me to exalt plants in the scale of organized beings; and I therefore felt an especial pleasure in showing how many and what admirably well adapted movements the tip of a root possesses."[8]

8 Charles Darwin: *Autobiography,* ed. Francis Darwin, p. 52.

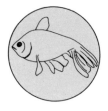

5. Harbingers of Modern Thought

More than half a century has now passed since Albert Einstein modestly announced that when matter is converted into energy, the amount of energy released is proportional to the square of the speed of light. Within less than thirty-five years, physicists confirmed this consequence of the special theory of relativity by measuring directly the amount of energy that is released when an atom splits.

In contrast, the field of biological rhythms has boasted no contemporary Einstein. In fact the first stirrings of thinking about how rhythms might be explained with any degree of rigor did not appear until the early 1930's. And when they did, the world of science greeted them with even more skepticism than it had accorded Einstein's theory—unbelievable as *that* had seemed in 1905.

Why should that have been so? There were a number of reasons, actually. For one thing, the living forms that biologists study are far more complex than the atoms of the physicists. The physicists' atoms, especially those of the heavier elements, are so complex in themselves that they have been aptly compared to a solar system in the universe. Nevertheless an atom is simple indeed compared to the smallest of living organisms. A virus particle so small that it can be seen only under an electron microscope—so small, indeed, that it may be called "living" only by definition—is still made up of a number of atoms so vast as to defy comprehension by any familiar yardstick.

For another thing, biologists in many different fields were attacking particular practical problems, and solving them; and since they found the answers that they were seeking, they went on to other things. As a result, considerable time had to pass before it became apparent that there were underlying common relationships among the rhythmic responses of a great many different kinds of organisms.

Entomologists, for example, ranged the world and found parasites to control crop-ravaging pests, but they did not notice that the adult insects emerged from their pupal cases at precise times of the day and year. Geneticists developed seeds better adapted to the latitudes in which they would be grown without consciously relating the length of the days to the season of the year. Animal pathologists learned the primary cause and found a cure for hog cholera long before it was realized that exposure to stress reduces a hog's resistance to the disease.

Thus it is not surprising that the rhythms that did attract attention had to be studied in detail by many different scientists before it came to be suspected that there might be some common relationship among them. And the concept of a living clock did not emerge until science had subjected a great many of these observations to the *synthetic process*—the logical process by which, in the words of John Dewey, men suddenly see that certain very different things all belong to the same *kind* of thing.

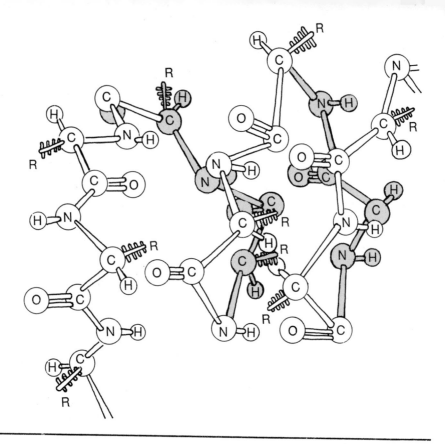

A very minute part of one molecule of one cell of a living organism. If the entire molecule were shown, it would extend more than 1,000 times the width of this drawing. The nucleus of one of the hydrogen atoms (H) would be too small to be visible on the scale of the drawing. The molecule shown is an alpha-helix of a polypeptide chain; indicated are hydrogen atoms (H), carbon (C), oxygen (O), and nitrogen (N). The symbol R represents complex radicals composed of carbon and hydrogen.

As examples, let us look at some pioneering experiments with organisms as different as jack beans, honeybees, and human beings.

One of the first precise analyses of a rhythm of a living organism was made by the Dutch botanist Anthonia Kleinhoonte in 1929. She was well aware of the work that had gone before, but she set out to break new trails.

She knew that Darwin had made hundreds of experiments on the sleep movements of leaves, but the experimental method had now advanced by fifty years, and she would have been inclined by then to class Darwin's results as "observations" rather than as true "experiments." His work had just not been planned to answer the kinds of questions that flooded her mind. Moreover, Darwin had made most of his observations either under natural conditions of alternating daylight and darkness or under simple variations of these conditions.

In particular, Darwin had made his observations on plants of more than one hundred fifty different genera, in some of which he observed the behavior of half a dozen different species. Thus, even if he had had the inclination, he would not have had the time to make a detailed study of any one of them.

Kleinhoonte, by contrast, felt that a deeper understanding of the sleep of leaves generally could best be achieved through the intensive study of the behavior of a single plant species. For her experimental material she chose the large jack bean, *Canavalia ensiformis,* and she laid out a carefully planned experimental program involving a very large number of measurements under a variety of different conditions.

The basic aim of her program was to learn as much as she could as to whether the sleep rhythms of her plants would appear to be inherited or would appear to be caused by the daily rhythm of alternating light and darkness. She asked herself what might happen if she took a plant that had been grown in normal conditions and then, at night, when the leaves would be in the sleeping position, she gave the plant a single flash of light lasting only a minute or two. What might happen if she grew plants from seed under abnormal cycles of light and darkness? Suppose she grew seedlings under a regime of six hours of light followed by six hours of darkness, repeated over and over? Suppose she tried eight hours of light and eight hours of darkness (LD 8:8 in today's laboratory shorthand), or LD 18:18, or LD 24:24? Would her seedlings learn such new rhythms? If they did, would such learned rhythms persist under constant conditions, such as continuous darkness, or continuous dim light?

Before she could begin to answer any of these questions, she needed a new kind of apparatus—equipment that would provide an accurate and continuous record of the movements of her bean leaves under any regime of light and darkness that she might wish to test. The method that Darwin had used obviously would not do, for that required continuous visual observation, and she did not wish to risk any possible upset that the dimmest of observation lamps might cause. She therefore decided on the kymograph, which was first developed in Germany and later improved by the Dutch botanist G. Brouwer. Now Anthonia Kleinhoonte, with a laboratory touch both imaginative and deft, built an exceptionally sensitive model.

When it was ready for its first trial, she attached a fine filament to the midrib of a primary bean leaf, and from that she passed a gossamer thread over a small pulley. Below the pulley she hung a piece of slender wire, with its lower end bent to form a stylus that would trace its path on a smoked drum rotating at a slow constant speed. The stylus was stabilized by a tiny weight suspended below it. A few trials showed that the device made beautiful records of the movements of the bean leaves. After each full rotation of the drum, Kleinhoonte changed the record, and by laying these out flat in succession from left to right, she had a continuous history of the leaf movement throughout many successive days.

To establish the best conditions for her controls, Kleinhoonte first sprouted bean seedlings under normal conditions. She found that the plants showed the daily rhythm of sleep best when they had reached an age of 17 or 18 days. Accordingly she selected plants at this age to make her first recordings and, attaching a primary leaf to the kymograph thread, she set the smoked drum in motion. Charles Darwin would surely have been fascinated had he been able to watch the smooth operation of the apparatus as it drew a continuous white line on the smoked drum.

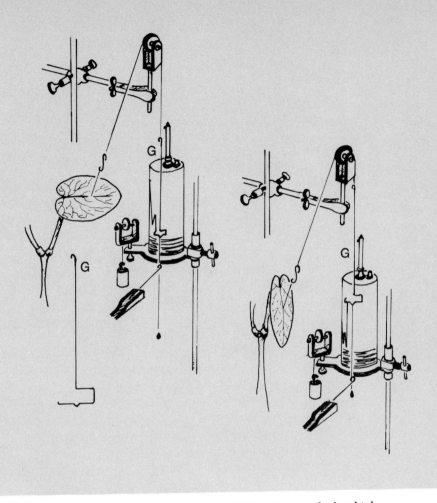

Anthonia Kleinhoonte used an aktograph like this to make her high-precision recordings of the sleep movements of bean leaves. (G) is the stylus that draws a line on the revolving drum to make the continuous record. The leaf is shown in the day position on the left, and in the night position on the right.

Once she had clearly established the pattern of sleep movement for seedlings grown under normal conditions, Kleinhoonte went on to test the abnormal conditions that she had laid out for her experimental program. First, she began recording the movement of a seedling under normal conditions, and then, during the night when the leaves were fully folded down in the sleeping position, she exposed the plant to a single flash of light of one minute. After that, she left the plant in continuous darkness. In that darkness, the plant returned to its regular daytime position (as the findings of earlier observers had led her to expect), *but the time when the plant resumed its rhythm was delayed by 12 hours.* Thus she showed that a single brief flash of light was able to reset the phase of the rhythm.

In her next experiment, Kleinhoonte put seeds to sprout in a completely abnormal light-dark cycle—a continuously repeating cycle of 8 hours of light and 8 hours of darkness. After the seedlings were 17 days old, she again attached the midrib of a primary leaf to her kymograph and set the smoked drum to rotating.

What would the record show? Would the plant behave as though the total length of the day were 16 hours, with 8 hours of "day" and 8 hours of "night"? That is precisely what it did. Again the phase had been reset by the abnormal condition. Then she made the critical test. She left the rephased plant under continuous light, so that the seedling no longer had any clue as to day or night, whether that was the abnormal one to which it had been set, or the natural one.

Would the plant now continue the abnormal LD 8:8 cycle? Or would it take up some other? She watched the stylus trace its path on the smoked drum. Gradually the plant shifted from the abnormal cycle to a perfectly normal daily rhythm that coincided with true day and night! Even though the plant had never "seen" true day and night from the instant it sprouted, even though it had been forced to accommodate itself to a completely unnatural cycle, it still had the capacity to take up a perfectly normal rhythm even though it had no external clue whatever.

In other experiments, Kleinhoonte used different abnormal cycles, and after the seedlings had come into phase with them, she exposed her plants to continuous light—in some cases— and to continuous darkness in others. And always the plants— once freed from the influence of unnatural cycles—returned to their natural rhythm in phase with the earth's daily rotation: the same rhythm to which the jack bean had become accustomed throughout countless generations.

Did this then mean that such a rhythm was truly inherited? Did it suggest that if jack beans were raised for many generations under constant laboratory conditions—and with no outside clue whatever as to the alternation of day and night—that the natural rhythm would persist? Anthonia Kleinhoonte indulged in no such speculation. She retained her objectivity. She said only that her experiments showed that periodic leaf movements were caused by "autonomic," that is, self-governing factors. She did not even speculate about what these factors might be.

Nevertheless, Kleinhoonte made a sharp advance in the imaginative design of her experiments, and her laboratory measurements were of amazing precision. Only through precision of this order could she have shown how precisely her plants responded to the apparent time of day. In the words of Erwin Bünning, who has been called the father of biological clock theory: "Above all, I should like to emphasize that the Dutch investigator Anthonia Kleinhoonte . . . made very decisive contributions to our understanding of the endogenous rhythm by the light-dark alternation."[1] Hers was a very long step forward leading to the first modern hypothesis about how living clocks work.

Another long step forward was taken by Ingeborg Beling, a graduate student working in the laboratory of the eminent Karl von Frisch, at the University of Munich. When Miss Beling first went to Munich in the late 1920's, von Frisch was already established as a world authority on communication among bees, and along with that he was also very much in-

---

[1] Erwin Bünning: "Opening Address: Biological Clocks," *Cold Spring Harbor Symposia on Quantitative Biology* (Cold Spring Harbor, L.I.: Long Island Biological Association, Biological Laboratory), Vol. XXV (1960), p. 2. Hereafter cited as *Cold Spring Harbor Symposia on Quantitative Biology*, with relevant article, volume, and date.

terested in how bees orient themselves in flight. The laboratory atmosphere that Miss Beling entered has been well voiced by von Frisch himself in his recently published reminiscences of that period:

"For more than 50 years," he wrote, "bees in our laboratory and during vacation times on the Wolfgangsee, have been the favorite animals of my scientific work. Their color vision, their smelling and tasting, and the relation of their senses to the world of flowers, their 'language' and their capabilities for orientation—this was the wonderland of puzzles that drew me ever onward. Little by little, much that is new has been revealed from the treasure chest of their being. These findings have mounted ever higher, and have become more and more insistent in their demands for contemplation and review."[2]

As early as the summer of 1912, a few months after he had achieved the coveted status of *Privatdozent* at the University of Munich, von Frisch had done his first important experiment with bees. At the family holding at Brunnwinkl on Lake Wolfgang he had shown that the bees from his hives could easily tell the difference between blue and various shades of gray. Once and for all, he had disposed of the theory that bees are color-blind. From this beginning, von Frisch went on to a lifetime of experimentation with bees. He developed and improved the design of observation hives, feeding stations for training, and methods of numbering bees with dots of colored shellac. By the latter method, he could instantly recognize any bee from No. 1 to No. 599; and by a minor variation of the system he could number as many as several thousand.

But more important than the excellence of the equipment for productive research was the boldness and imaginative quality of von Frisch's experimental designs, and the stimulating suggestions that he made to his graduate students. Thus to Munich came a steady stream of gifted young people, anxious to work under the direction of the renowned von Frisch.

[2] Karl von Frisch: *The Dance Language and Orientation of Bees* (Cambridge, Mass.: Harvard University Press; 1967), p. v.

Among them, Ingeborg Beling was to make one of the most lasting contributions.

In discussing possible areas of research that she might carry out, von Frisch was reminded of Forel's observation that bees came to the breakfast table on his terrace in Switzerland, whether or not sweets were on the table; and of von Buttel-Reepen's report that bees came to the buckwheat fields only during the hours when nectar flowed from the blossoms.

Here was a most challenging problem, as the professor pointed out to Miss Beling. Both Forel and von Buttel-Reepen were sure that the bees came for food only at the particular times when they might expect it to be there. More than that, von Buttel-Reepen had actually called this remarkable ability a "time sense." Now what was the true source of this time sense? Does the bee have it inside herself, or does she get clues from the outside world? If it is clues from the outside, just what events give rise to these clues?

Ingeborg Beling at once saw the fascination of the problem, and eagerly agreed to take it up. She was already skilled in taking care of observation hives, and with practice she would be at home with the special experimental techniques that von Frisch had developed over the years. With youthful enthusiasm she plunged into planning her research program. That enthusiasm she was to need.

The first phase that she planned was to determine the exact behavior of the bees under specific conditions in a perfectly normal environment. For this she trained numbered nectar-collecting bees to come to a feeding station supplied with sugar water, at a particular time each day. In a typical experiment, she used bees numbered 1, 11, 12, 17, and 19, and trained them to come to the feeding station between four and six o'clock in the afternoon. Let us watch as she prepares to number five bees so that she can identify them at a glance in the field.

Before her, on a laboratory table, is a rack containing five fine brushes dipped in alcohol to keep them soft, and five small culture dishes containing shellac previously colored white, red, blue, yellow, and green. Close at hand is a watch glass containing sugar solution, and around its rim five bees cluster, happily drinking. Miss Beling dips her first brush into the white shellac dish and places a small dot at the middle of the thorax nearest the head of the No. 1 bee. On bee No. 11 she places two white dots, also close to the head. Bee No. 12 receives a white dot at the upper left and a red dot immediately to the right of that. Bee No. 17 gets a white dot toward its head and a yellow one below it. Finally, No. 19 gets a white dot above and a yellow one below. Throughout the process, all five bees have been so busy drinking the artificial nectar that they have paid no attention at all to Miss Beling's activities. After the shellac has dried thoroughly, she returns the bees to their hive with small net tongs. The next step is to train the bees to go to a particular feeding station during a single two-hour interval each day.

Professor von Frisch had already worked out just how bees could be trained to do this, and now Miss Beling began with her own cluster of five bees. The steps were like those for training any other animal—that is, by very gradual progression.

Early one morning, before any bees had started abroad, Miss Beling placed a small drop of sugar solution just at the entrance to the hive. Especially important was the fact that the solution was scented with the faintest trace of oil of lavender. The bees soon found it and with that had taken the first step in the routine that they were to learn.

The next morning she placed another drop of "nectar" a bit farther out. The third morning, she used her feeding table. This was simply a small slab of wood mounted on a stick that had been sharpened on the opposite end, so that she could shove it into the ground wherever she wished. She now put the stake ten feet away from the hive, and on the table she placed a watch glass containing scented nectar. Even though the food was now ten feet from the hive, the bees soon found it. They had begun to learn the direction of flight that they were being trained to take. She was then able to move the feeding table farther and farther from the hive, and soon the bees knew the fixed location that Miss Beling had chosen for a main part of the experiment.

Now that her bees had learned the correct direction of flight, Miss Beling's next step was to train them to come for food only at a particular two-hour interval during the day. For this case, she chose the hours from 4:00 to 6:00 P.M. Only during those hours was nectar on the feeding table; at all other times the watch glass was empty. Even more rapidly than they had learned the proper direction of flight, the bees learned that it was a waste of time to go to the feeding table at any time other than late in the afternoon. The next step was to make an exact record of what each marked bee did during all of the foraging hours of a day when nectar would be withheld throughout the entire day.

On the morning of July 20, 1927, Ingeborg Beling rose early. By 6:30 A.M. she was already at her observation post, notebook, pencil, and watch at hand, ready to write down the arrival time and bee number for every visit. More, she was prepared to maintain her vigil without a moment's interruption until the last visit of the day—probably some time after eight o'clock that evening.

Looking forward to a long and eventless watch from half past six in the morning until four in the afternoon, Miss Beling remained keenly aware of the need to stay alert. To her surprise she was reminded of this before an hour had passed. The drone of a single foraging bee caught her ear. Landing on the feeding table the bee approached the empty watch glass.

She saw two white spots on the anterior part of the thorax. That was bee No. 11. It seemed strangely early for any bee. Even more strange, No. 11 came back a second time between seven-thirty and eight o'clock that morning.

Days later, when Miss Beling was discussing the results of the experiment with Professor von Frisch, she asked him why he thought a well-trained bee should get so far off schedule. From his own long experience he told her that she should always keep in mind that biology is not an exact science. There are always occasional sports in behavior, just as there are deviations from the norm in other areas. Was not this particular case comprehensible biologically? As he had once written: "Better be early than late at the feeding place, in a world full of hungry creatures."

But for the rest of that morning of July 20, 1927, no other bee visited the feeding table. In fact, no other forager came until half past three. Between then and four, No. 11 was back again, and also No. 19. In the next half hour, 6 bees came, and between four-thirty and five, there were 17 visits. After that, during each half-hour, there were 11 visits, 4 visits, and 2 visits. By half past six, the quest was over, and although Ingeborg Beling maintained her watch until after eight, no other bee came. Then she rose wearily and made her way back to the laboratory, carrying her precious record of just what each bee had done throughout that day and exactly when she had done it.

She also had a somewhat uncomfortable memory of how her bees had behaved when they found that there was no food in its accustomed place. As Max Renner, a later researcher described it: "On a test day, the observer might judge by their behavior that the bees cannot 'believe' that the table has not been laid for them as usual. Over and over again, they come flying, run around searchingly, stretch out their proboscises into the bowl to reassure themselves that there is really nothing sweet in it. The most industrious ones among them extend their searches after a while to the wider environment and lick all glittering objects—such as the watch and pencil case of the observer—and some especially audacious ones will even poke their proboscises into the wrinkles of the note-taking observer's bent hand."[3]

Ingeborg Beling's long day did much more than confirm earlier observations that trained bees will visit an empty feeding station at an appointed time. It was actually the beginning of an extensive accumulation of data that she could use as a reference standard. Once she had established that standard, she could test her bees under any unusual conditions

[3] Max Renner: "The Clock of the Bees," *Natural History*, Vol. LXVIII (October 1959), p. 434.

that she might choose and know how much the change of conditions changed the behavior of the bees. She would know, that is, how much information about the normal diurnal cycle the bees needed to be able to find a given place at a given time.

Therefore, on many other test days, Miss Beling gathered more data, using other times of the day for her test period, and larger numbers of bees. Altogether, it is an impressive collection of numbers. For Miss Beling, it set the stage for the crux of her program.

What events in the outside world, Professor von Frisch had asked, might give the bees their clues?

What, then, were some known cyclical events? Miss Beling listed them. Alternation of light and darkness, obviously. Changes in the position of the sun in the sky. Variations in atmospheric temperature, with warm days and cool nights. Variations in the electrical charge of the atmosphere, as had been noted decades earlier by Arrhenius. Variations in cosmic radiation. And finally, some unknown factor or factors—some subtle geophysical force that, though not perceived by man, could still give a time-clue to a bee. Such a force is not hard to imagine. On a more mundane level, we know that a dog can hear an ultrasonic whistle that is inaudible to a man.

Miss Beling saw a feasible way to eliminate variations in light level, temperature, humidity, and electrical charge in the atmosphere. The facilities of the Zoological Institute of Munich University provided her with a specially designed room in which she could keep each of these factors at a constant level. And she found after many, many experiments that her bees performed just as regularly in conditions of constant light, constant temperature, constant humidity, and constant atmospheric charge as they had outdoors on a sunny July afternoon.

One of her co-workers at Munich, O. Wahl, also showed that variations in cosmic radiation were not the time-giving factor. He took his bees six hundred feet below the surface of the earth, into the gallery of a salt mine, and trained them successfully there.

Perhaps, then, the bees did not need any outside clue at

all. If they were trained to some period quite different from the earth's 24-hour cycle, the memory of that training period might produce a continuation of that cycle. To test this, Beling tried to train bees to respond to a 19-hour cycle; and in this, she failed completely. Strong evidence, indeed, that the clock of the bees is regulated by some unknown external factor that is in phase with the earth's 24-hour rhythm.

What might that factor be?

The search for it turned to a series of attempts to understand the nature of the clock itself. If this were known more clearly, then it might be possible to make a more fruitful guess as to what regulated it. The first attempt to do this, based on the questions that Anthonia Kleinhoonte and Ingeborg Beling had left unanswered, was undertaken by Erwin Bünning in 1936.

Meanwhile, however, an American psychologist who had become interested in rhythms in human physiology was developing a theory of a "chemical clock." Developing it, indeed, while quite unaware of the contemporary work in Europe[4]— just as those in Europe were unaware of what he was doing.[5]

In 1931 Hudson Hoagland took his new Harvard Ph.D. in psychology to Clark University in Worcester, Massachusetts, where he began a life-long study of the relations between the physiology of the human body and the activities of the human mind. For a number of years he was professor of general physiology and chairman of the Biology Department at Clark; in 1944 he became executive director of the Worcester Foundation for Experimental Biology, and after many years of service, was named its President Emeritus.

Under a board of distinguished editors he was the author of the first volume of a pioneering series of monographs on experimental biology. In that monograph he stressed the view

---

[4] Hudson Hoagland to Ritchie R. Ward, personal communication, January 15, 1970.
[5] Erwin Bünning to Ritchie R. Ward, personal communication, January 23, 1970.

that "much of the overt behavior of organisms is determined by the interrelations between chemical events within cells and groups of cells, quite independently of external environmental factors."[6]

This approach would therefore be called "mechanistic," that is, subscribing to the view that the behavior of living organisms can be fully explained by the laws of physics and chemistry. Further, Hoagland was one of the first to believe that the study of rhythmic processes offered a basic key to understanding human behavior. He was especially impressed by Archibald V. Hill's description of two different kinds of oscillation processes that are found in nature and by Hill's use of one of these as a model in his studies of nerve tissue. At the first Cold Spring Harbor Symposium on Quantitative Biology in 1933 Hill said:

> Whenever we look in the world of matter and events outside ourselves we find that oscillations and wave motions have a significant, often a dominant role. It is not, therefore, astonishing to find that waves play an important part in ourselves also.
>
> Most of the well-known oscillations with which physics is concerned are the consequence of the reaction with one another of properties analogous to inertia and elasticity. A moving or a changing system tends, on the one hand, to continue in its state of motion because it possesses, for example, mass or inductance; even social, economic and intellectual changes are endowed with such characteristics of inertia, which keep them going when they have passed a true condition of equilibrium. On the other hand, such systems, if they are to continue to exist, if they are not merely dissipated, must possess converse properties which tend to bring them back once they have overshot their equilibrium: such properties of physics are elasticity and electrical capacity, in finance and politics are fear and conservatism.[7] These exercise a constraining force increasing

---

[6] Hudson Hoagland: *Pacemakers in Relation to Aspects of Behavior* (New York: The Macmillan Company; 1935), p. 1.

[7] A generation later, William X. Scheinman wrote of the stock market: "There is a rhythm to successful investing as surely as the ebb and flow of the oceans. Investment timing involves the matching of one's major

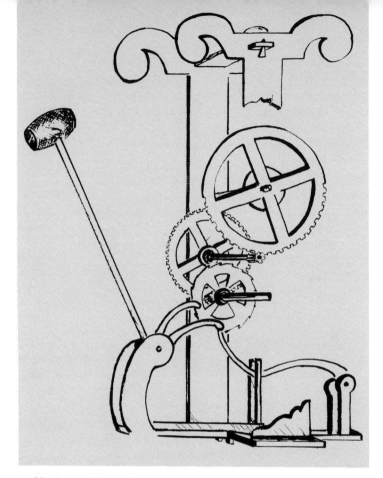

Galileo's pendulum clock, an example of an inertial oscillator. The pendulum periodically overshoots the equilibrium (vertical) position to the right and to the left. At the points of maximum overshoot the constraining force of gravity dominates, and the direction of motion is reversed.

with the displacement from equilibrium, and ultimately reverse the motion or change, and the same oscillation is repeated in the opposite direction.

decisions to the basic rhythms. But the rhythms are not always simple ones. The basic melody of the market is joined by the counterpoint of the individual stocks. . . . These waves represent the alternating waves of investor optimism and pessimism." (William X. Scheinman: *Why Most Investors are Mostly Wrong Most of the Time* [New York: The Scheinman Divergence Analysis Report Corporation; 1968], p. 89.)

There is, however, another type of oscillation, less commonly discussed in physics or mechanics, but nonetheless well known in everyday affairs; that which depends upon a discharge taking place when some limiting potential or intensity is reached. For example, (1) water falling into a tank equipped with a syphon will come out in rushes whenever it rises to a certain level. Or again (2) a population in which an epidemic of measles cannot start because of the number of people in it who are immune, having had the disease already, will gradually become less immune as time goes on, and finally an epidemic, a wave of measles, will sweep through it. (3) A neon lamp with a parallel condenser, in series with a resistance and a source of electromotive force, will discharge at regular intervals, namely, whenever the potential difference across the condenser reaches a critical value. This type of oscillation does not depend upon inertia reacting with elasticity. Its essential nature is (a) that some state, some potential, some intensity, is built up by a continuous process, and the condition becomes less and less stable until one is reached in which discharge must take place, and (b) this discharge once started forms a path for itself by which (as in a syphon or an electric arc) further discharge is facilitated until what has been built up gradually has been broken down, and the process begins again. This type of oscillator (sometimes referred to as a relaxation oscillator) is the one with which we alone are concerned in physiology.[8]

Following Hill's lead, and as a physiologist also, Hoagland naturally centered his interest on oscillators of the relaxation type. And since he thought that chemical explanations of them were most likely to be fruitful, he was continually on the

[8] Archibald V. Hill: "Wave Transmission as the Basis of Nerve Activity," *Surface Phenomena*, in *Cold Spring Harbor Symposia on Quantitative Biology*, Vol. I (1933), p. 146.
    Archibald V. Hill was a British physiologist who was awarded the Nobel prize for medicine and physiology jointly with Otto Meyerhof "for his discovery relating to the production of heat in muscles." Although he is best known for that work, he chose the foregoing entirely different subject for his paper at the first Cold Spring Harbor Symposium on Quantitative Biology. The concept of the relaxation oscillator that he developed there is relied on today in the most advanced speculations about how the biological clock works.

lookout for processes that would be expected to change in the
same way as the rates of chemical reactions change. Two important influences on chemical reaction rates are temperature
and the concentration of the substances taking part in the reaction.[9] Hoagland was able to collect experimental evidence on

[9] Simple chemical reactions double their rate (approximately) for each
10° rise on the Centigrade scale. Complex reactions, such as boiling
an egg, often deviate significantly from this rule.

*A Greek water clock, an example of a relaxation oscillator. When
the uniform flow of water from the cylinder at the left raises the
water level in the cylinder at the right to its maximum head, water
is discharged through the syphon, and as the angel atop the float
sinks, it writes a descending line on the revolving cylinder; the
cylinder is turned by the waterwheel.*

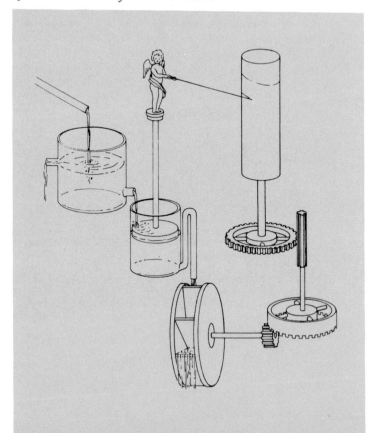

both effects from human beings in ordinary situations. His own wife was the subject of one of his studies on the effects of temperature.

> A . . . likely basis for a subjective time scale would appear to be . . . a chemical mechanism, perhaps associated with the continuous respiration of cortical nervous tissues. With this hypothesis in mind, experiments were made to determine the effect of internal body temperature on judgments of short durations.
>
> The idea of making these experiments came essentially as a product of adversity. It resulted from the fact that my wife, having fallen ill with influenza, was impressed with the fact that time seemed to pass very slowly. It occurred to me that this might be due to an elevated temperature from fever since, if some form of chemical reaction acted as a timing mechanism the increased body temperature should make this chemical clock run faster. This should make time, as judged by objective clocks . . . appear to go more slowly, since more psychophysiological "time" would pass in a given constant unit (a minute, let us say), of our conventional constant time scale standard.
>
> Without in any way hinting the nature of the experiment, I asked my wife to count 60 to herself at a rate of what she believed to be one per second. Simultaneously the actual duration of the count was observed with a stop watch. The reciprocal of the number of *actual* seconds occupied by the count was taken as a direct measure of the rate of the physiological process underlying the count (i.e., the reciprocal of time has the dimension of rate). These data, when plotted according to the Arrhenius equation gave [a straight line].[1]

That is to say, the higher her temperature, the more slowly *(in direct proportion)* time seemed to Mrs. Hoagland to pass.

Dr. Hoagland had also noted that changes in the concentration of hydrogen ions, as well as of carbon dioxide, in the tissues of the respiratory center (such as are produced by mild exercise) speed up one's breathing rate, and speculated that "it

---

[1] Hoagland: *Pacemakers*, p. 107.

is not unreasonable to suppose that chemical changes locally produced in the brain . . . would decrease the velocity of our hypothetical 'clock,' making time appear to pass more rapidly when we are active."[2]

From such observations, Hoagland drew the conclusion that "physiological time . . . seems to depend directly upon the velocities of certain definite chemical processes, the psychological and physiological events forming different aspects of the same thing."[3]

Clearly, then, Hoagland thought that most, if not all, internal clocks varied with temperature (as well as with concentrations of certain chemical substances in body tissues). However, we have already seen that a clock that is not independent of temperature would be of little use to its owner, and we are faced with an apparent inconsistency.

Fortunately, however, the inconsistency is only apparent. When we look more closely, we see that the temperature differences that Hoagland dealt with are quite small (a few degrees only), while the temperature range over which an organism needs to have its clock running accurately may range from below zero to over a hundred degrees in the shade.

The notable advances that Hoagland made were these: He was one of the first to suggest that living beings have a chemical clock—a "pacemaker"; and he was surely the first to see clearly the very wide field over which the study of clocks might be extended with rewarding results. "From all this it appears that the further elucidation of kinetic mechanisms . . . should be a desirable objective for students of the behavior of living organisms," he concluded, "whether these students call themselves biochemists, biophysicists, physiologists, botanists, zoologists, behaviorists, or psychologists."[4]

We shall see that as time went on, workers in amazingly diverse fields did indeed use the clock concept as a hypothesis.

[2] *Ibid.*, p. 113.
[3] *Ibid.*, p. 119.
[4] *Ibid.*, p. 123.

Karl C. Hamner inspects his apparatus for recording bird activity. To gain deeper insight into the flowering of soybeans, he needed to study the rhythms of house finches.

Moreover, we shall see that researchers on rhythms had to familiarize themselves with a far wider range of specialties than most scientists ever encounter as a professional necessity.

Thus Karl Hamner, a California botanist, found that he must study the rhythms of birds to understand better the behavior of soybeans. R. G. Lindberg, a specialist in radioactive fallout for the Atomic Energy Commission, began recording the periodic behavior of desert pocket mice.

But consideration of these is ahead of the story. Just now, we visit the Arlington Farm of the U.S. Department of Agriculture as it was in 1920, and where the Pentagon stands today. There two scientists, who were looking for a better method of growing tobacco, made the first discovery that a living organism can actually measure the length of a period of time.

6. Garner and Allard: Short-Day Plants and Long-Day Plants

In the temperate regions of the world the climate changes with the changing seasons—often dramatically. In winter in northern latitudes the mercury may drop to forty degrees below zero, the snow drift higher than a man's head, and the leaden skies foreshorten even the briefest days. To survive such rigors, plants form seeds and tubers, insects enter rest stages, and hibernating animals turn dormant. The adaptive significance of such behavior is obvious.

Not so obvious, perhaps, is the fact that living organisms anticipate the arrival of winter well before it comes. Seeds cannot form overnight, and plants prepare to produce them long before the snow flies. How can they know when the proper time has come?

Two ways will come to mind. Organisms either measure

the passage of time in a calendar sense (the poinsettia somehow realizes when it is December), or they sense the shortening
of the days as winter approaches. In the latter case, they
would only need to know the length of the day to know what
time of year it was.

This is a most important point, and before we go on, we
must make a further clarification. In all of our discussions so far,
we have spoken of the time sense of living organisms quite
generally. Thus we have often used the term "living clock" to
indicate nothing more than that some kind of repetitive cycle
of events in time was occurring. Now, however, we must be
more precise. A rhythm is not necessarily a clock.

An analogy of Erwin Bünning's will help to make the
point clear. If we drive a stake into level, sunny ground, we
have the first element of a sundial. But it is not yet a clock.
It can only become a clock after we have marked off the positions at which the shadow falls at each hour of the day.
Similarly, many of the phenomena that we have been concerned with have been rhythms that were indeed in phase with
the solar days, but just from that we could not infer that the
organisms were using the rhythms to tell time. In fact the
first report that any living organism can measure time was not
published until 1920.

That report, written by W. W. Garner and H. A. Allard, of
the Bureau of Plant Industry of the U.S. Department of Agriculture, is a landmark.[1] The authors, however, did not begin
this work out of any desire to explore the time sense of living
organisms; what they set out to find was a better way to
propagate tobacco.

It was already well known that the Maryland Mammoth, a
new tobacco variety, produced excellent leaf in profusion
(some plants yielded more than one hundred leaves), but
refused to flower before frost in the Maryland-Virginia area,

---

[1] W. W. Garner and H. A. Allard: "Effect of the Relative Length of
Day and Night and Other Factors of the Environment on Growth and
Reproduction in Plants," *Journal of Agricultural Research*, Vol. XVIII
(1920), pp. 553–606.

and farmers could not obtain seeds from it when it was grown in the field in the ordinary way.

Garner, a chemist turned physiologist, and Allard, a plant physiologist and naturalist with a wide knowledge of both wild and cultivated plants, took up the problem of how to obtain seed from this very desirable tobacco variety. Garner was Physiologist-in-Charge of the Office of Tobacco and Plant Nutrition Investigations, and, while much of his time was taken up by administrative duties, he was especially interested in the problem of the Maryland Mammoth. Often he visited the greenhouses and fields where Allard was carrying out the experimental work; most days, too, they lunched together, discussing current progress and outlining new experiments.

Allard afterward recalled the period in a letter to Paul de Kruif.[2] For his immediate research problem, he had been asked to find out why the Maryland Mammoth flowered only late in the season, so that seeds could not form before frost killed the plants. Accordingly in 1918 he began with a study of the hereditary and growth characteristics of a number of varieties of tobacco, including, of course, the Mammoth.

A related problem also intrigued the two scientists. As Garner later wrote: "The two cardinal observations with which we started were that, in contrast with other tobaccos, the new Maryland Mammoth variety always continued a purely vegetative type of growth through the open growing season in Washington, and that successive plantings of certain varieties of soybeans made at short intervals through the spring and early summer all tended to flower at the same date. In the case of tobacco, it was not at first realized that a seasonal effect was definitely involved in the very unusual behavior of the new variety."[3]

[2] H. A. Allard to Paul de Kruif, personal communication, January 19, 1931. I was able to see a copy of this long and historically valuable letter through the courtesy of the Southern Historical Collection of the University of North Carolina Library.

[3] W. W. Garner, personal communication, cited by A. E. Murneek: *Vernalization and Photoperiodism* (Waltham, Mass.: Chronica Botanica Company; 1944), p. 41.

Photoperiodic control of flowering. Above, *Maryland Mammoth tobacco in eight-inch pots exposed to light from 9:00* A.M. *to 4:00* P.M. *daily near Washington, D.C. The seed pods were full grown on August 15, 1919. Below, control series of Maryland Mammoth plants kept out of doors. There were no signs of flowering on August 15.*

Thus the problem was defined, and Garner and Allard began their attack on it. And, during two years of intensive research, they met discouragement after discouragement. They would set up a hypothesis that seemed reasonable, test it experimentally through the tedious, time-consuming trials that plant studies demand, and find that their results contradicted the hypothesis. They would thrash through the problem again, and hammer out another hypothesis. And that would fail.

This failure is hardly surprising when we consider the state of knowledge about the flowering and fruiting of plants as it stood at the close of World War I. Garner and Allard summarized the views of the day in a paper which they also wrote in 1920 for publication in the *Agricultural Yearbook:*

"One of the most characteristic features of plant growth outside the tropics," they said, "is the marked tendency shown by various species to flower and fruit only at certain periods of the year. This behavior is so constant that certain plants come to be closely identified with each of the seasons, in the same way as the coming and going of migratory birds in spring and fall. In midwinter the blossoms of cyclamen, freesia, the brilliant color of poinsettia, and the fruits or berries of ardisia, all are reminders of the season; in spring we expect to see the unfolding blossoms of forsythia, wild violet, crocus, redbud, dogwood, and other typical plants; as summer approaches, poppy, rhododendron, iris, and columbine begin flowering; in the autumn salvia, aster, cosmos, dahlia, and chrysanthemums herald the approaching end of the open growing season."[4]

Garner and Allard then went on to discuss the possible reasons that might lie behind these seasonal displays. "The

[4] W. W. Garner and H. A. Allard: "Flowering and Fruiting of Plants as Controlled by the Length of the Day," *Yearbook of the Department of Agriculture, 1920* (Washington: U.S. Government Printing Office; 1920), pp. 377–400. All references to Garner and Allard in this chapter pertain to this article. Today's reader, knowing that he can have dahlias, for example, in midsummer (or at Christmas if he lives in a mild climate), may be surprised at this emphasis on the seasonal nature of flowering. Of course the explanation is that in the fifty years since 1920 the blooming periods of many traditionally seasonal flowers have been much extended— in some cases through new varieties and in others through improved cultural practices.

thought at once suggests itself that the underlying cause or causes of flowering or fruiting occurring only at a particular season must be purely internal, else the vagaries of the weather and other variable external conditions would seriously upset the regular cycle. It is true, of course, that plants can flower and fruit successfully only within certain limits of temperature and moisture supply, and it has long been known, also, that light is indispensable. Thus plant development may be retarded in the spring by cool weather, and at times drought or excessive rainfall may interfere, but in general flower and fruit are produced regularly in their seasons in spite of these temporary disturbances. The ripening of seeds as a sequel to flowering is obviously of great importance to many plants, in that it affords the only means of avoiding extermination. We might easily conclude from this that the plant's entire activities are directed toward this means of propagation, all preliminary growth and development of root, stem, and leaf being incidental. This view, however, is not correct. The plant merely inherits the capacity to flower and fruit in response to certain favorable external conditions."

In the light of today's knowledge, this was indeed a prophetic statement. Note that although the authors seem to be weighing a choice between internal and external control of the time of flowering, they recognize—at least implicitly—that internal processes and external influences might be working hand in hand: "The plant merely *inherits the capacity* to flower and fruit in response to certain favorable *external conditions.*"

With this, they foreshadowed by a decade one of the most original and fruitful ideas in all of biological rhythm theory, as we shall see when we come to explore the Bünning hypothesis.

But such considerations were only incidental to the problem that Garner and Allard had set for themselves, and as they thought further about which external conditions might be involved, they also pondered the effects of geographical location on plant behavior: "While marked regularity in the time of flowering and fruiting is the rule in plants, so long as they are grown in any particular locality in temperate regions, transferring plants from one region to another may greatly change their habits. A species which flowers and fruits readily in one region may become sterile in another, or, in some instances, the time of flowering may be changed from spring to fall, or vice versa. Again, plants behaving as annuals in one region may become biennials in another. These changes in the behavior of plants when grown outside their native regions furnish strong evidence that external conditions control the processes of flowering and fruiting, and also suggest the possibility of artificial control. . . . It is both interesting and practically important, therefore, to determine these conditions."

Now they set out to identify these conditions. Recall Garner's granting that they did not realize at first that the behavior of the Maryland Mammoth involved a seasonal effect. From the results of their early experiments in the greenhouse, they suspected that the shock of transplanting to the greenhouse might be a major factor, and further that this effect might be accentuated by the increasing age of the plants.

They were also "temporarily misled by the observation that seedlings grown in small pots in winter and early spring flowered and fruited freely, thus suggesting a nutritional angle. The big lead came after several years with the observation that with the advance of spring new shoots developing on stumps that had been producing flowering shoots suddenly swung over to the indeterminate vegetative type of growth. It was then perfectly clear that a seasonal factor was involved."[5]

What, then, might that seasonal factor be? In their report for the 1920 *Agricultural Yearbook* they first considered temperature:

We instinctively think of temperature as the outstanding factor causing one season to differ from another in its effects on plants. In particular, we associate the opening of spring flowers with moderate temperatures, following the chill of winter. Likewise, as the characteristic flowers of autumn make their appearance we have been inclined to assign decrease in tem-

[5] Garner and Allard: "Flowering and Fruiting."

perature as the cause; mainly perhaps for the reason that there has seemed to be no other obvious cause for the flowering of these plants. Temperature unquestionably is a very important factor in plant development, and plants differ widely in their temperature requirements. Nevertheless, change in temperature fails to explain why plants flower and fruit at certain periods; that is to say, even though the appropriate temperatures are provided out of the regular flowering and fruiting season, as a rule the flower and fruit fail to appear except in their usual seasons. For example, common iris, which flowers in May and June, will not blossom under ordinary conditions when grown in the greenhouse in winter, even under the same temperature conditions that prevail in early summer. Again, one variety of soybeans will regularly begin to flower in June of each year, a second variety in July, and a third in August, when all are planted at the same date. There are no temperature differences during the summer months which could explain these differences in time of flowering; and since "internal causes" alone cannot be accepted as furnishing a satisfactory explanation, some external factor other than temperature must be responsible. . . .

Certain varieties of soybeans change their behavior in a peculiar manner with advance of the summer season. The variety known as Biloxi, for example, when planted early in the spring in the latitude of Washington, D. C., continues to grow throughout the summer, flowering in September. The plants maintain growth without flowering for 15 to 18 weeks, attaining a height of 5 feet or more. As the dates of successive plantings are moved forward through the months of June and July, however, there is a marked tendency for the plants to cut short the period of growth which precedes flowering. This means, of course, that there is a tendency to flower at approximately the same time of year regardless of the time of planting. As a necessary consequence, the size of the plants at the time of flowering is reduced in proportion to the delay in planting. . . . [Thus] the Biloxi soybeans show a marked tendency to flower at a definite season of the year, and if planted early they wait, as it were, till this season arrives. It is easy to see the advantage which a plant has in being able to shorten the growing period which must precede flowering if, for any reason, the plant gets a late start. In such a case the chances of successfully maturing the seed before frost and thus avoiding extermination in a given region are greatly increased, and the production of seed constitutes the plant's method of perpetuating itself in the face of the destructive action of the cold. It is important, however, to make a distinction between advantage and cause with respect to time of flowering. The Biloxi soybeans by curtailing the period of vegetative activity when beginning growth late in the season are actually able to forestall the arrival of cold weather; hence low temperature cannot be considered as a cause of this behavior. The response of the soybeans to the advance of the season begins before there is any decrease in temperature.

In the letter cited earlier, Garner stressed this same point: "Since the plants were growing in a warm greenhouse, it appeared that temperature could be excluded." Then he noted other variables that he and Allard had taken into consideration: "With respect to light, we naturally reasoned that intensity and composition might be involved. However, fairly extensive investigations were being conducted on other problems with both tobacco and soybeans that seemed to preclude intensity and spectral composition of light as important factors."

And just at that point Garner made a deeply revealing statement—revealing as to how he felt at the time, and revealing as to what often happens in the course of an arduous research program. Just on the eve of the discovery that "led to a great upsurge in related research,"[6] Garner had despaired: "At this stage the problem appeared to assume a somewhat hopeless trend, but after much deliberation it was concluded that the only remaining seasonal phenomenon that could be a factor was change in relative length of day and night. The importance of this conclusion, of course, lies in the fact that length of day was dissociated from the factor of amount of solar radiation. The decisive test was then made in the simplest possible way

[6] Karl C. Hamner: "Experimental Evidence for the Biological Clock," *The Voices of Time*, ed. J. T. Fraser (New York: George Braziller, Inc.; 1966), p. 282.

but without great hope of success, mainly for the reason that there appeared to be no accepted basis in plant physiology for consideration of day length as other than a purely quantitative factor."

Allard, in a letter of his own, gave further details about the inception of the idea, and then told the immediate circumstances of the key discovery.[7]

The clue came to both scientists at the same time, Allard recalled. In an informal and rambling exchange of ideas, they went over a number of factors that might induce flowering. As they reviewed Allard's most recent experiments, they thought again of conditions that vary with the season, such as temperature and light; also quite casually, they mentioned the length of the day as a possible influence. Unlikely as that seemed, Allard devised a simple way to test it. He built what he called a primitive dog house, which was ventilated, and in which he could maintain plants in total darkness. Thus he could expose experimental plants to the same shortened length of "day" as though it were late fall. And in July of 1918 he put a few buckets of Maryland Mammoth tobacco into his dog house for 14 hours every day, and brought them into the daylight for 10 hours. Soon these experimental plants set flowers, while control plants, fully exposed to the long days of summer in the field, remained vegetative until fall.

This, then, was the homely way in which Garner and Allard discovered that plants can measure the length of the day, and from that, sense the season of the year. Allard wanted to call the effect simply the "length-of-day response," but that did not sound sufficiently scientific to the administrators of the United States Department of Agriculture, and somewhere along the line the term "photoperiodism" was coined.[8] In any case, by the time they came to write the conclusion to their first paper, Garner and Allard themselves said: "The term *photoperiod* is suggested to designate the favorable length of day for each organism, and *photoperiodism* is suggested to desig-

[7] Allard to de Kruif, personal communication, January 19, 1931.
[8] *Ibid.*

nate the response of organisms to the relative length of day and night."[9]

Allard's brief description of the discovery of photoperiodism makes it sound simple; of course it was far from that. That first test—encouraging as it was—was no more than a lead. A long road lay ahead before the two scientists were ready to write their paper for the 1920 *Agricultural Yearbook*. Let us follow them along that road.

At Washington, D.C., the time between sunrise and sunset ranges from nearly 15 hours in late June to about 9½ hours at Christmas. To determine whether this change in the length of day is a cause of regularity in the time of flowering and fruiting, a series of experiments was made in which a number of plants were darkened for a portion of the day during the long days of summer. The results obtained were remarkable. The plants no longer persisted in their usual habit of deferring the flowering period till a particular time of year had been reached. The normal seasonal periodicity was completely broken up.

The experiments included a large variety of plants both wild and cultivated, and it was found that the reaction to differences in the length of day is of very wide occurrence.

The method followed in these tests is very simple. A "dark house" was so constructed as to admit air freely at the bottom and allow its escape at the top, without the admission of daylight. For convenience a series of small steel tracks leading into the dark house was provided, and on these tracks were mounted a number of trucks with steel wheels capable of supporting the containers in which the plants were grown. With this

[9] Garner and Allard: "Flowering and Fruiting." Since the original suggestion, the term "photoperiodism" has came to subsume a very wide range of phenomena. By 1957, James Bonner of the California Institute of Technology told a conference of experts: "A great many facts about photoperiodism have accumulated since the formulation of the principle by Garner and Allard in 1920. These facts are not only numerous but complex, and it does not appear to be possible [even] today to put all of them together into one coherent model." (James Bonner: "The Photoperiodic Process," *Photoperiodism and Related Phenomena in Plants and Animals*, ed. Robert B. Withrow [Washington, D.C.: American Association for the Advancement of Science; 1959], p. 245.)

equipment it was a simple matter to transfer the plants into and out of the dark house at regular intervals each day. For example, if it were desired to give a particular lot of plants eight hours of light each day the truck bearing these plants would be rolled into the dark house at, say, 4 o'clock in the afternoon each day and rolled out into the open air again at 8 o'clock the following morning. For comparison, in each test a second lot of plants, known as "control" plants, was grown under exactly the same conditions as those to which the darkened plants were exposed, except that the control plants were exposed to light throughout the day.

The response of the plants to this artificial shortening of the daylight period was prompt and clean cut. Biloxi soybeans which germinated May 17 were allowed to receive seven hours of light daily, beginning May 20. These plants were in blossom in 26 days, whereas a similar lot of plants exposed to light throughout the day required 110 days to flower. This variety of soybeans, which ordinarily flowers in September, even though planted in May, was forced into blossom in June, simply by shortening the daylight period. In further tests it was found that a daylight period of 12 hours was as effective as the 7-hour period in forcing the flowering of the soybeans. It is easily seen, therefore, why this variety of soybeans ordinarily does not flower till September, for it is at that time that the length of the day is reduced to 12 hours.

An experiment was made with another variety of soybeans known as Peking at the same time and in the same way as with the Biloxi. In this case the plants receiving 7 hours of light flowered in 21 days, while those exposed to light for the entire day required 62 days to reach the blossoming stage. This is fully in accord with the fact that the Peking regularly blossoms in the field in July, two months in advance of the Biloxi. The Peking, therefore, is capable of flowering under a considerably longer day than the maximum day length which will cause the Biloxi to blossom.

A common wild aster which ordinarily flowers in September was found to behave in the same manner as the Biloxi soybeans when exposed to a shortened daylight period. When exposed to 7 hours of light daily the aster was in bloom in 36 days, as against 122 days when exposed to light for the entire day. A variety of lima bean imported from Peru which ordinarily does not flower till late in the fall at Washington, D.C., was caused to blossom in 28 days by reducing the daily light period to 7 hours. The common ragweed behaved in a similar manner.

One scarcely expects to see chrysanthemums in bloom in midsummer . . . but these typically fall-flowering plants are readily made to flower in summer by shortening the length of the daily light period. Late-flowering varieties of dahlia are readily forced into blossom during the summer by reducing the length of the daily light exposure to 10 hours or less. A highly colored specimen of poinsettia, the plant so typical of the Christmas season, was developed in August by reducing the daily light period to 10 hours.

In the light of these experiments there is no longer any element of mystery concerning the fact that when plantings of cosmos are made at successive dates in early spring a point is reached at which the plantings suddenly swing over from flowering in the spring to flowering in the fall. Cosmos begin to flower in the fall when the length of day has decreased to about 12 hours (sunrise to sunset) and, in the same way, it is no longer able to flower in the spring after the days become much in excess of 12 hours in length.

There is, then, a large group of plants, including most of the so-called summer annuals, which regularly flower after midsummer as a result of decrease in the length of day. While relatively short days favor flowering and fruiting in these plants, long days are more favorable to rapid and extensive vegetative development. Some of these plants, therefore, if they receive the full benefit of the long days of summer, may reach giant proportions before being brought into the flowering condition. Thus, we can understand why it is that when the farmer plants some crops too early, there is a tendency toward excessive development of leaf and stem with little flowering and fruiting. Late planting, on the other hand, may lead to dwarfing in growth but abundant flowering and fruiting. Again, it is easily seen why carrying some plants into northern latitudes causes very rank growth, with a tendency toward barrenness, since the length of the day in summer increases as we go northward. Plants in this group differ widely as to the extent to which the longest summer days must be shortened to induce flowering, with the result that some flower in July, while others may not flower till November. Even the latest of these are readily forced into flowering and fruiting during the hottest part of the summer simply by shortening the daylight period, so that there is no reason for considering the cooler weather of fall as a factor of importance.

People who work in specialized fields develop shorthand terms for ideas that take many words to express; scientists, especially, excel at this. Soon, then, Garner and Allard began calling "varieties of plants that require short daily periods of exposure to light to reach the flowering or fruiting stage" simply "short-day plants." The converse immediately became obvious, and from their rich store of knowledge of the habits of wild and cultivated plants, Garner and Allard began to see many varieties that could clearly be classified as "long-day plants." Other varieties fell into an intermediate group whose members were quite indifferent to the length of the day.

Their report continues.

In striking contrast with the group of plants already discussed is a second group regularly flowering in late spring and early summer. It is obvious that these plants do not require short days to reach the flowering stage. On the contrary, it has been found that short days prevent, or at least greatly delay, flowering and fruiting. To this class of plants belong the so-called winter annuals; also many of our common vegetables. The radish has given some interesting results which are fairly typical for the group. The ordinary varieties of radish when planted in the spring first produce a thickened edible root and somewhat later develop a flowering stem, which in due season matures seed. Thus the Scarlet Globe variety, planted May 15, began to blossom June 21 when exposed to the natural length of day. A similar planting, made at the same time, but allowed to receive only 7 hours of light daily, grew slowly and formed no flowering stem. Under the shortened daylight period the roots of the radishes continued to enlarge slowly throughout the summer, with a corresponding increase of the rosettes of leaves surmounting the roots. One of the plants which was transferred to the greenhouse in the fall continued its slow growth through the winter months. Finally, as the days lengthened in early spring this plant was able to send up a flowering stem and perished after seed formation was completed. Thus the radish, which ordinarily is a typical annual, was made to behave as a biennial. The radish furnishes a case in which flowering can be prevented for more or less indefinite periods by shortening the daily period of illumination, in contrast to the group of plants previously considered, which are prevented from flowering by long days and are forced into flowering by shortening the daylight period.

The behavior of the radish is in no sense exceptional. Failure to send up a flowering stem during the short days of winter and early spring is a characteristic feature of many hardy plants which maintain more or less vegetative activity at those seasons of the year. The tendency is toward a prostrate type of growth, with free stooling [putting forth shoots from the base of the plant], or a rosette form of leaf development. As the longer days of spring come on, the character of growth changes, and upright-growing stems appear, in preparation for flowering and fruiting.

Our small grains belong to this class of plants. Red clover furnishes a good illustration of this behavior. . . . By allowing the test plants to receive only 10 hours of light daily, the prostrate nonflowering type of development was continued long after a corresponding lot of plants which were exposed to light all day had developed upright stems and had successfully flowered and fruited. Likewise, the common evening primrose transplanted from the field in early spring continued the prostrate rosette type of development for several weeks when allowed a daylight period of only 10 hours, whereas similar plants exposed to light throughout the day quickly developed tall, erect flowering stems.

Under ordinary conditions spinach can not be grown successfully for table use during the summer months, because it quickly goes to seed instead of forming the desired rosette of large leaves. This behavior has been generally attributed to high temperature. It is quite true that within suitable limits an increase in temperature, as a rule, speeds up plant development. Nevertheless, experiments have shown that spinach will produce an excellent rosette in summer if the light period is reduced to 8 or 10 hours. Under these conditions the flowering

External view of the Canberra phytotron of the Commonwealth Scientific and Industrial Organization in Australia.

An interior view of the Canberra phytotron shows many small cabinets with independent day-length and temperature controls. The entire building works on the heat pump principle.

stems are unable to form, or, at least, their appearance is greatly delayed. . . .

The above examples illustrate the fact that there is a large group of plants which are brought into the flowering and fruiting stages of development because of the increase in length of day as spring advances into summer. As a matter of convenience in discussing flowering and fruiting activities, this group may be spoken of as "long-day plants," in contrast with the group previously discussed, which are forced into flowering and fruiting by the shortening of the days in fall and therefore may be called "short-day plants."

While as a whole there are sharp contrasts between the two groups, there are many plants which perhaps may be regarded as occupying an intermediate position. There is, in fact, no hard and fast line between these two classes of plants. There are some plants, indeed, for which it is possible to provide a daylight period too long, on the one hand, and too short, on the other, to induce flowering and fruiting.

Of course Garner and Allard could easily cut the amount of light that they allowed their short-day plants simply by wheeling them into the dark house at any time they chose. Conversely, to give their long-day plants more light during short days they could obviously use electric lights; that gave them no particular problem, but they did need to determine what levels of light would be effective.

In summer the daily light period is readily shortened by use of dark chambers into which the plants are placed for a portion of the day. In this way various plants may be forced into flowering and fruiting out of their natural season, or plants normally flowering and fruiting in summer may be prevented from doing so. On the other hand, to initiate flowering out of season in long-day plants during the short days of winter, or prevent its occurrence in short-day plants, it would be necessary to lengthen the daily period of illumination. With this in view, a greenhouse was fitted with a series of 40-watt electric lights, evenly distributed overhead, so that an average intensity of about 3 to 5 candlepower was obtained immediately above the soil surface. The electric light was used from sunset till about midnight each day. The intensity of the light used seems insignificant in comparison with daylight, which on clear days in winter may reach as high as 5,000 foot-candles or more.[1] Yet some striking results were obtained. For comparison, plants were grown in a similar greenhouse without the use of electric light.

As a general proposition, the long-day plants should tend to remain in the purely vegetative condition in the "control" house without electric light, and hasten toward reproductive activity in the electrically lighted house. Short-day plants, on the contrary, should flower readily in the control house and assume a purely vegetative form in the the illuminated house. In the control house cosmos has invariably flowered, showing reproductive tendencies when very small. Flowering actually took place within 50 to 60 days from germination. In the illuminated house the plants grew vigorously, greatly exceeding the control plants in stature, and showed no indication of flowering, months after the controls had flowered. These plants were removed from the illuminated greenhouse in June and placed out of doors, where they received only the normal daylight of the long summer days. Under these conditions the plants remained in the actively growing, sterile, vegetative stage and did not flower till they had reached a height of 15 feet in October, when they were finally forced into the reproductive stage by the natural decrease in day length. . . .

These examples are enough to show that artificial light of low intensity used to prolong the daily illumination period during the short days of winter effectively prevents many short-day plants from flowering, and is equally effective in forcing long-day plants into flowering and fruiting. Comparatively weak artificial light used as a supplement to daylight of short dura-

---

[1] Nearly twenty years later (1938) Hamner and Bonner made an observation that was to throw a completely new light on time measurement in photoperiodism: they found that the effects of a long dark period could be nullified by a very brief light signal given in the middle of the dark period. (Karl C. Hamner: "Endogenous Rhythms in Controlled Environments," *Environmental Control of Plant Growth* [New York: Academic Press, Inc., 1963], p. 215.) "A short day naturally contains a long dark period, but it was found that such a short day would be interpreted as a long day if the long dark period were interrupted for a very short time with but a few foot-candles of light. This discovery led many botanists to believe that the responses were simply to the length of the dark period." (Karl C. Hamner: "Experimental Evidence for the Biological Clock," p. 283.)

tion during the winter will produce much the same effects as the daylight of long duration in summer.

To close their paper in the 1920 *Agricultural Yearbook* Garner and Allard speculated on the leads for further research that were opened by the discovery of photoperiodism, and discussed the improvements in cultural methods that could be put to use at once.

The experiments . . . discussed in this paper have opened a wide field for experimentation and study. The full significance of the discovery that the activities of plants are profoundly influenced by seasonal change in the length of day cannot be understood until the field has been more fully explored. At present it is possible only to indicate broadly some of the directions in which it seems most likely that practical application of the principles involved can be made.

A correct interpretation of the effects of length of day upon the plants will be a great aid in reaching a better understanding of the causes which limit the natural habitat of most plants, a problem which has been a difficult one to solve. To the farmer, the facts which have been established will strongly emphasize the importance of accurately knowing the correct season for planting each of his crops in order to secure the highest returns. Under some conditions a difference of no more than 10 days in time of planting will definitely direct the plant's activities toward either the purely vegetative or the reproductive form of development, as the case may be. Now, in one case the farmer may be chiefly concerned with extensive vegetative growth, while in another he may be primarily interested in flower, fruit, or seed development. Of course, much has already been learned empirically as to the proper time of planting various crops, but recognition of the importance of the relative length of day and night as a factor in a measure reopens the question.

The plant breeder should be able to gain a better insight into some of his problems, such as securing for any particular region earlier or later varieties, more fruitful or larger-growing forms, and improved everbloomers and everbearers. In the same way the problem of extending the northern or southern ranges of crop plants may be more clearly defined. In many cases breeding work can be hastened through artificial control of light duration, which will make it possible to work more or less independently of natural conditions of day length, both as to time of year and as to geographical location of the worker. It often happens that plant breeders are unable to make crosses between certain plants because of differences in time of flowering of the two parental types. In such instances artificial control of the daily light period should be of great value, for in this way the date of flowering can be accurately controlled. The plant introducer will have at his command a more adequate basis for analyzing the factors which determine whether any particular plant is adapted to a new region. Moreover, in special cases it may be possible to introduce successfully new plants through artificial control of light conditions or by taking fuller advantage of seasonal differences in length of day.

Within suitable limits of temperature and other important factors in plant growth, there would seem to be no reason why almost any plant may not be made to flower and fruit at any season of the year and in any region. By shortening the daily light period through the use of dark chambers or lengthening it by means of artificial light, reproductive activities may be induced almost at will. With proper knowledge of the specific requirements of each plant, therefore, the florist should be able to force flowering at any desired time of the year. It has been possible to secure excellent flowering specimens of iris in midwinter, and chrysanthemums, poinsettias, and other plants in summer by utilizing these principles. In the same way wild violets have been kept in the everblooming stage as long as 9 months. The principles involved are so simple that anyone interested in plants can easily obtain instructive and convincing results.

What then of the problem that Garner and Allard set out to solve in the first place? How did they get seed from the Maryland Mammoth tobacco? Their solution was simple and direct: "The problem of securing seed is easily met by growing the plants in southern Florida during the winter, for under these conditions the Mammoth flowers and fruits much the same as the ordinary varieties. . . ."

Thus they met the immediate need. They also made available a powerful new technique by which farmers, flower growers, and plant hybridizers could control the timing of their crops with a precision never before possible. The impact of the discovery on practical horticulture was great, and application of the principles spread fast.

But it is a striking curiosity that ten years were to pass before the true significance for theoretical biology of the discovery of photoperiodism was recognized. It was as though no one apprehended the demonstration that plants can tell time. Not until 1930 did Erwin Bünning start laying the groundwork for modern theory about how living clocks work.

7. A Clock-Watcher's Dictionary[1]

The history of science teems with misunderstandings, controversies, animosities, and disputes, both between individual
scientists and between opposing schools of thought. Some of
these have arisen when the issues were clear-cut, and the
parties simply agreed to disagree. Many, however, have been
due to simple misunderstanding, and these have come about
most often when no uniform terminology was available.

Exchange of ideas about biological rhythms suffered for
many years for lack of a common language, and indeed it was

---

[1] A large body of experimental data on biological rhythms was obtained
without the aid of automatic recording equipment, and the researchers
who obtained it often kept watch for twenty-four or more unbroken
hours. Tempering fatigue with humor, they often referred to themselves
as "clock-watchers."

not until 1964 that a uniform terminology was proposed. In September of that year, sixty-eight scientists from thirteen countries gathered at Feldafing, West Germany, under the sponsorship of the NATO Scientific Affairs Division, to exchange ideas about living clocks. Earlier symposia had seen violent controversies, and the organizers of the Feldafing conference were sure they would see more of the same unless something was done to lift the burden of confused terminology. Accordingly three authorities—Jürgen Aschoff, Karl Klotter, and Rütger Wever—prepared a list of terms and their definitions that has since become standard.[2]

Originally intended for specialists, these definitions have been expressed with exceptional clarity by Professor Beatrice M. Sweeney, a distinguished authority on plant rhythms. In doing this, Professor Sweeney added illustrations of experimental strategies that have added much to the versatility of present-day rhythm research. We are privileged to turn to this now.

To relate the definitions to specific plant and animal behavior, I have interpolated examples drawn from other chapters in this book; these comments appear in square brackets.

A SHORT DICTIONARY FOR STUDENTS OF RHYTHMS[3]

In any field there are certain ideas which occur very frequently. Sometimes the expression of these ideas in everyday language requires many words and awkward explanatory phrases. For this reason, a special vocabulary soon develops, a convenient shorthand which may be obscure to the uninitiated. In the field of rhythms this is particularly true, since the recurrent ideas are somewhat more than ordinarily difficult to express in common parlance. Many observations are neces-

[2] Jürgen Aschoff, Karl Klotter, and Rütger Wever: *Circadian Clocks* (Amsterdam: North-Holland; 1965), p. x.
[3] From Beatrice M. Sweeney: *Rhythmic Phenomena in Plants* (New York: Academic Press, Inc.; 1969). Used by permission of the author and the publisher.

sary to demonstrate the presence of a rhythm and one often wishes to describe all these data together in the form of a graph, with time as the abscissa. [The abscissa is the horizontal scale of a two-dimensional rectangular graph.] Then, too, it is often necessary to refer to the light conditions where a light period of, say, twelve hours alternates with a dark period of the same or of a different length, and is followed by constant light. Consequently, shorthand expressions have been developed to describe both the conditions of rhythm experiments and their results, thought of graphically.

Perhaps the first thing to define, so that we shall be quite clear what we are talking about, is the term "biological rhythm." A rhythm is a regular fluctuation in something, e.g., in the position of a leaf [like the leaves of de Mairan's sensitive plant, moving up and down], or in the rate of a physiological process, photosynthesis for example, or in shape, in the production of spores, in color, in activity. The important thing is that a *repeating* pattern can be discerned clearly. However, every repeating pattern is not a rhythm. A rhythm is self-sustaining and it is important to understand what this implies. Perhaps an example will make this clear. Let us say that we observe that the leaves of a tree in the garden are folded together every night but spread wide again every morning. There is a repeating and predictable pattern in this behavior, but is it a biological rhythm? We cannot tell, since the enormous difference between night and day, which is itself a repeating pattern, may be *causing* the differences which we see in the leaves. They may open in light and close in darkness. If this is the case, the change in leaf position is not a biological rhythm but simply a response to light, and the rhythmic pattern of the leaf position is not self-sustaining, and would disappear at once were the environmental changes removed. Just this experiment must be done to detect the presence of a true rhythm. It must be shown to be self-sustaining, at least for a time, in constant conditions of light and temperature. [The level of light in de Mairan's closet was constant, because he kept his sensitive plant in continuous darkness; and we may assume that the temperature was also quite constant. Under these conditions the leaves of the sensitive plant continued to move up and down at dawn and at dusk, and the rhythm was therefore self-sustaining.]

Before we consider the terms which are used to describe rhythmic phenomena, let us pause to think about a repeating pattern with which we are familiar, the turning of a wheel. The name "cycle" comes from this analogy where a circle is changing its position in time, yet any point on the rim is always returning to contact the ground. There are several ways in which we can picture what is happening to points on this wheel, as it turns. From a vantage point to the side of the direction in which the wheel is rolling, a point on the rim will appear to follow a scalloped course (see below), while looking at the wheel from in front, the same point will describe a sine

Derivation of cycles from the turning of a wheel. From a vantage point to the side of the direction in which the wheel is rolling, a point on the rim will appear to follow a scalloped course (a), while looking at the wheel from in front, the same point will describe a sine curve (b).

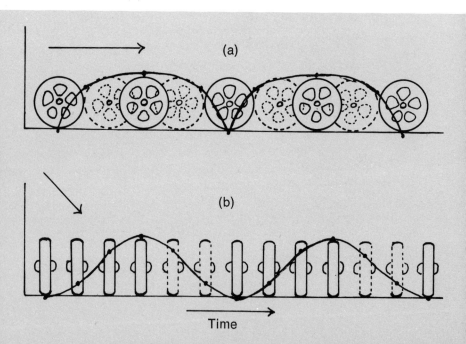

(a)

(b)

Time

curve. [A sine curve is the graphical representation of the equation $y = \sin x$, $\sin x$ being the ratio of the two sides of a right triangle that contains the angle x. Such a sine wave is generated, for example, when the onset of activity of a nocturnal animal comes later and later as summer approaches, because the sun sets later and later. Thus Patricia DeCoursey's flying squirrels (page 207) began running at 4:30 P.M. in January, and at 7:30 P.M. in June. The form of the curve throughout an entire year was sinusoidal.] These curves are cyclic and come, as we see, from a circle in motion in time. Their pattern repeats over and over. It is an "oscillation." The fundamentally oscillatory character of rhythms makes it appropriate to use for them the same terms used in describing physical oscillations like that of the wheel, the pendulum, or the electronic oscillators. The time required for an oscillation to make a complete cycle and return again to the starting position is the "period." In the wheel analogy, it is the time for one rotation of the wheel. In the rhythm in leaf movement, where leaves are spread open maximally every day at noon, the period is the time from noon to noon, or 24 hours. Note, though, that the period is the same no matter what point in the cycle is used to measure it. [When we show the distance that leaves move during the sleep-wake cycle over a short span of days, the effect of changing daylength is minor, and the curve assumes the form of a square wave, as shown—simplified—in the figure on page 140. Regardless of the shape of the curve, the period is still the time required for a complete cycle. The period indicated on the right of the figure extends from one dawn to the next. Under normal conditions the period of a plant's sleep rhythm is just 24 hours. But when de Candolle kept sensitive plants in *continuous* light, the periods were reduced to 22 hours (page 50).]

As in physical oscillations, the inverse of the period is the "frequency," the number of cycles which take place in a given time. [Thus the frequency of de Candolle's sensitive plant in normal conditions was 1/24; and in continuous light it was 1/22.] In the discussion of biological rhythms, frequency is used less often than period, however. Just as in physical oscillations, the frequency of rhythms may be forced to match

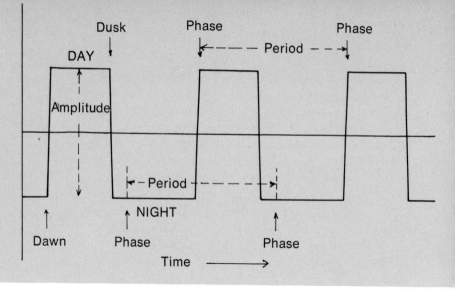

"Dawn-and-dusk," or square wave, illustrating three terms used to describe oscillations.

exactly that of some external oscillation, the alternation of light and darkness for example. When this occurs, the rhythm is said to be "entrained" by this external oscillation. The signal responsible for entrainment has been called the "zeitgeber" or "time-giver." [The similarity to Hudson Hoagland's "pace-maker" (pages 104 ff.) is obvious.] Entrainment, of course, can only take place while the external oscillation is actually present. When the rhythm is *not* entrained, it is said to be "free-running," and now shows its *natural* period. [De Candolle's sensitive plant was free-running in continuous dim light, and its natural period was 22 hours.]

Under certain special conditions, oscillations can be entrained to show periods which are a multiple of the entraining cycle. This phenomenon is called "frequency demultiplication" since the entrained period which the rhythm shows is longer than that of the external cycle which is entraining it. [If a plant is exposed to an abnormal cycle of 6 hours of light and 6 hours of darkness, over and over, it responds just as though it were exposed to a 12-hour light:12-hour dark cycle. Erwin Bünning says that "this is not surprising considering the fact that only short periods within the physiological 24-hour cycle are particularly sensitive to light." Since in the example the entraining cycle is 6:6 and the resulting entrained cycle is 12:12, the phenomenon is called frequency demultiplication.]

"Amplitude" is the term used to describe the intensity of an oscillation. [In the figure opposite, the amplitude of the rhythm is labeled at the left. The amplitude of the leaf movement of Darwin's Virginia tobacco was the distance between the lower and the upper dots in the figure on page 87.] A rhythm with a large amplitude is one in which there is a big difference between the maximum and the minimum in the quantity which is showing the rhythmic variations.

In discussing a process that is changing in a rhythmic manner, not only do we need a term for the time that it takes to make one complete cycle and return to the starting point (the period), but it is necessary also to describe just where in this cycle the rhythm is at any certain time. Is the process at its maximum, minimum, ascending or descending in rate? For the stage of the cycle the term "phase" is used. [Two different phases have been labeled in the figure opposite.] This term is often employed to describe where the cycle is relative to some other time basis, say sidereal or solar time. Two rhythms may be said to be out of phase when their maxima do not coincide. It is also possible, as we shall see, to change the phase of rhythms in various ways. This means simply changing the solar time at which some phase occurs, moving the maximum from noon to midnight for instance. Since oscillations are really projections of circular motion—as we saw in the analogy above with the rotating wheel—the phase may be described in degrees, and so the phase of one rhythm relative to another may be measured in degrees, as "90° out of phase." Changes in phase are also measured in this way; a change which results in a maximum where a minimum formerly occurred is then a 180° phase shift, or half of a circle. Thus the phase may be changed by moving the whole cycle forward or backward in time, *without changing the period.* Changing the phase of a rhythm is also called "resetting" the rhythm by analogy with moving the hands of a clock. When alterations in phase are produced, it sometimes happens that the complete change does not occur in a single cycle, but may take several cycles before the final new phase is established. During this transition, the apparent period will be either shorter or longer than usual and will be unstable, perhaps changing with each cycle. Such intermediate cycles are known as "transients."

When the new maximum occurs earlier than the old, the phase is said to be changed in the positive direction, while delay in the maximum is regarded as negative. Curves in which the change in phase is plotted as a function of when in the cycle a single stimulus was applied are called "response curves," and values for the amount of phase shift are either positive or negative according to this convention. [A phase response curve for the rhythms of the luminescent one-celled plant *Gonyaulax polyedra* is shown in the figure opposite.]

In physical systems, oscillations usually take on a wave form. Although some biological rhythms also appear to follow a wave pattern, this is by no means universal. Whatever the apparent shape of the biological oscillation, however, the terms period, phase, and entrainment are used.

Biological systems which oscillate may be classified in various ways. Biological rhythms in the true sense of the word are only those oscillations which can be shown to continue in the absence of periodic changes in the environment, such as changes in light and temperature, i.e. self-sustaining oscillations. Such rhythms may be called "endogenous" since they apparently arise from inside the organism. Aschoff, Klotter, and Wever refer to them as *"active systems,"* since they are self-sustaining oscillations capable of deriving the energy necessary for their maintenance from constant sources of energy. This is in contrast to "passive systems" which appear rhythmic by virtue of their ability to respond to periodic variables in the environment, and cannot derive energy for their continuation from any constant source. Such periodicities are also called "exogenous," since their origin is outside the cell. They are not true rhythms, since they are not self-sustaining. [The process of photosynthesis in plants, for example, is periodic; it begins at dawn and stops at nightfall,

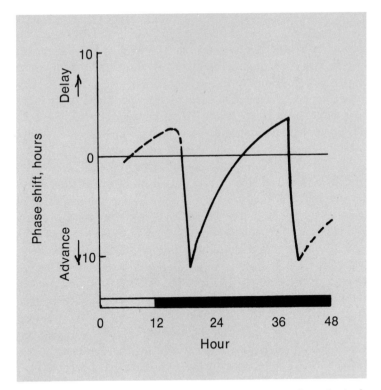

A phase response curve of the rhythms in Gonyaulax. A single three-hour exposure to light shifts the phases of the glow rhythm, the stimulated luminescence rhythm, and the rhythm of cell division.

day in and day out. But it is not self-sustaining; it cannot continue in the dark.] Since many true rhythms gradually die out or "damp out" in constant conditions, it is not always easy to distinguish a true rhythm from one of these exogenous periodicities, especially since we are not always sure that the environment does not contain some periodic variable to which the organism can respond, perhaps one of which we, the experimenters, are unaware.

Biological rhythms may also be classified according to the length of the period they show, as "diurnal" or "circadian" [the term "circadian" was first suggested by Franz Halberg, who derived it from two Latin words meaning "about a day"], when the period is about 24 hours, "tidal" when the period approaches 12.8 hours, "lunar" if 28 days and "semi-lunar" if 14–15 days, "annual" if one year.

The most frequently used of the shorthand expressions that have been devised are those which describe the conditions of light or darkness used in experiments on rhythms. Alternating light and darkness is represented by L(light), D(darkness), and the number of hours of each is written directly afterward, light duration first, then darkness, both in hours. A light-dark schedule in which 12 hours of light is followed by 12 hours of darkness and this routine is repeated over and over is written as LD 12:12 for example, or if the light period is 16 hours and the dark 8, as LD 16:8. The sum of hours of light and darkness need not be 24, of course, as for example LD 8:8. [In the foregoing example of frequency demultiplication, the conditions of the entraining cycle were LD 6:6.] The light intensity in "lux" [one lux is the illumination produced by a light of one international candle falling at right angles onto a surface one meter away from it] may be written in brackets after the hours of light and darkness, as LD 24:24 (5,000). In this notation, LL 100 designates continuous light of 100 lux, and DD, continuous darkness.

These are the principal terms which we will need to describe rhythms. With tools thus properly sharpened, we are now ready to tackle the subject of what is known of these rhythms in biological systems, and what is still to be discovered.

8. The Bünning Hypothesis

In the summer of 1928 Erwin Bünning, slender, fair, intense, entered the office of the director of the Institute for the Study of the Physical Basis of Medicine at the University of Frankfurt am Main, Germany. Young Bünning was just finishing his studies in botany at the University of Göttingen. There he had given much attention to the irritability of plants, and he certainly was puzzled as to why he was being asked to consider a possible appointment to a medical institute.

The director of the institute, Professor Friedrich Dessauer, introduced a second young man of about Bünning's own age, Kurt Stern, who was also a doctoral candidate in botany. Professor Dessauer, amused at the mystified faces before him, went on to explain why he wanted botanists for the new research posts at his institute, rather than physicians or physiologists.

The great Swedish chemist Svante Arrhenius, Dessauer said, had published a paper some thirty years before in which he had reported a relationship between cosmic forces in the environment and human diseases. In particular he had found that bronchitis and epilepsy were related to variations in the electrical charges of the atmosphere, or, as Dessauer put it, to variations in the ionic content of the atmosphere. If that were true, it might well point to ways of relieving the effects of these diseases.

But no one had pursued Arrhenius's observation any further, and it seemed to Dessauer that the Frankfurt Institute for the Study of the Physical Basis of Medicine was in an ideal position to do so. He was also convinced, however, that human subjects would be far too complex, physiologically, to start research with. As the first subject, some simple organism seemed more promising, and he had wondered if a carefully selected plant species might not provide a suitable research tool. He wanted, in short, a plant that would show some sort of observable response to changes in the ionic content of the air. And hence his search for two research-minded young botanists.

Erwin Bünning had hesitated at first, for he was full of plans for further research on the irritability of plants. But as he thought more about it, he saw that if plants really were affected by electrical charges, that would be a kind of irritation, and perhaps this very problem would provide a stimulating new approach to the whole field of plant irritability. So Bünning accepted Dessauer's offer with eager anticipation.

Kurt Stern agreed also, and when the 1928 fall term opened at the University of Frankfurt, the two young botanists were hard at work.

Their first task was to make an intensive survey of the literature, to find out what plants might be suitable for the kinds of experiments that Dessauer wanted, and how their reaction to ions in the air might be observed.

Before long they hit on work that Rose Stöppel had done at Hamburg some years earlier. She had rechecked some of Wilhelm Pfeffer's work on plant rhythms, and had selected the calendula as the plant with which to work. She had been able to show more convincingly than Pfeffer that the leaves took the same position at exactly the same time each day, because her observations were more precise, and she had a much larger body of data.

From the great precision that the calendula leaves showed, she concluded that the rhythm could not arise within the plant itself, for if it did, every plant could not be expected to behave so much like every other; surely there would be individual differences from plant to plant. Surely some of them, at least, would drift out of phase from clock time. But not one of them drifted, and Rose Stöppel concluded that there must be some subtle "factor X," as she called it, that was responsible for the precise time response.

That was just the clue that Bünning and Stern were looking for. Here was a plant whose leaf movements could be measured and recorded very precisely, and which responded, Rose Stöppel thought, to some subtle "factor X" in the environment. Was it possible that that factor might be the atmospheric ionic content of Svante Arrhenius? The two young botanists were eager to find out.

The excitement of new discoveries lying ahead charged the atmosphere in the laboratory. Quickly Bünning and Stern assembled the necessary apparatus. Meanwhile they set out seeds of scarlet runner beans to germinate, for they thought that plant would be more convenient to work with than Rose Stöppel's calendula.

When their young plants had grown large enough for testing, they would attach a fine thread to a leaf blade, and fasten the other end of it to a lever, which, in turn, would move a pen across a revolving drum—an arrangement similar to the aktographs that previous researchers had used. The critical difference from Stöppel's procedure was that they placed a single plant inside a light-tight box, with the thread passing through a light lock, so that the plant inside was completely shielded from light.

To start with, they repeated Stöppel's routines to confirm for themselves the nature of the leaf rhythm, and to check out the operation of their own apparatus. All was in good order, and they began at once a series of experiments in which they artificially varied the ionic content of the air in the dark box. As they watched the pen trace off the leaf movements on the revolving drum they were at first puzzled, and then worried, and finally deeply disappointed. Something was very wrong.

Feeling totally inadequate, they reported their results to Dessauer. The concentration of the ions in the air had had no effect whatever on the bean seedlings. They had done their experiments over and over. They had increased the ionic content of the air in the dark box much more than ever occurs in nature. But in every case the rhythms of the leaves remained unaffected. The electrical charge in the atmosphere was certainly not factor X.

Dessauer reassured the two young scientists. Of course he was disappointed. Naturally he had hoped that the plant studies would open up a new approach to understanding human diseases. But if that did not lead to the truth, then they would pursue the truth in whatever direction it might seem to lie.

Erwin Bünning and Kurt Stern trudged back to the laboratory. With meticulous care they repeated the experiments of Stöppel, over and over. And as they did so, they noticed a small difference begin to appear on their records. As their experiments continued day after day, the bean leaves slowly drifted out of phase from clock time! One plant in continuous darkness gradually shifted to a free-running rhythm of 22 hours. Another lost time, and finally settled down to a rhythm of 27 hours.

How could this be, when Stöppel's plants had always maintained a precise 24-hour rhythm? Was there after all a mysterious factor X that operated in Hamburg but not in Frankfurt? But that was too ridiculous!

As with a fine tooth comb the young botanists doggedly went through Stöppel's description of her procedures, step by careful step, looking for some small difference, however insignificant it might seem.

They had known all along that Stöppel had kept her plants in a darkroom, rather than inside a light-tight box like theirs. Each day at a regular time she had gone into the darkroom to water her plants. Bünning and Stern had provided means for watering their plants without opening the dark box.

Could this apparently insignificant difference be what they were looking for? How had she been able to see to do her watering? In her paper she had said that she had used a dim red light, and that she turned it on when she went into the darkroom.

But that should not have affected her plants. All plant physiologists knew that plants will not turn toward a red light as flowers turn to the sun. Knowing this, how could they believe that a brief exposure to a dim red light would have any effect on leaf movement?

Nevertheless, that was the only difference they could find between Stöppel's procedures and their own.

These fine imaginative scientists could not disregard the challenge here. Could a dim red light have some effect not previously observed? Should they try a red light in their dark box? Ridiculous.

Yet futile as it seemed, they did try it. They put a red light in the dark box, and each day they turned it on at exactly the same time, as Stöppel had done.

The plants responded at once.

Instead of drifting out of phase with the clock, *they maintained the exact time that was signaled by the lighting of the red light.* Here was the mysterious, subtle factor X! Nothing more than the red light that Stöppel had turned on when she watered her plants!

So Rose Stöppel had guessed right in the very beginning.

Plant rhythms *did* drift out of phase from the earth's daily rhythm when they were allowed to run free under genuinely constant conditions. The rhythm must be truly endogenous. If different plants under constant conditions took up periods different from the earth's, and also different from plant to plant, then factor X was no longer mysterious. It did not even exist.[1]

Having thus finished his work for Professor Dessauer, Erwin Bünning returned to the University of Berlin and completed his doctorate. This he received in 1929—anything but a good time for a young botanist to be seeking an academic post in Germany.

Of the possibilities that offered themselves, Bünning centered his interest on those that would give him freedom to continue his research on plant rhythms, for during his work at Frankfurt, he had lost interest in plant irritability in favor of the fascinating ramifications of rhythms.

He decided on the Botanical Institute at the University of Jena, in what is now East Germany. The name "Institute" suggests much more in the way of facilities than were available there in 1930. When Bünning arrived, the institute had one professor, one associate professor, one lecturer besides himself, and a housekeeper and his wife who, besides their regular duties, did the cleaning of the institute building. The total budget for teaching and research (exclusive of salaries) was one thousand dollars, and that allowed no clerical or typing help; in fact, not even a telephone.

Still, the young lecturer was not unhappy with his post. His regular teaching and administrative duties required his full attention only five days a week from eight to twelve in the morning, and two to six in the afternoon; and on Saturday mornings from eight to twelve. All other times were free for research. He had a salary of forty dollars a month, and he had funding for all of his research equipment up to twenty-five dollars.

[1] F. B. Salisbury and .C. W. Ross: *Plant Physiology* (Belmont, Calif.: Wadsworth Publishing Co., Inc.; 1969), p. 535.

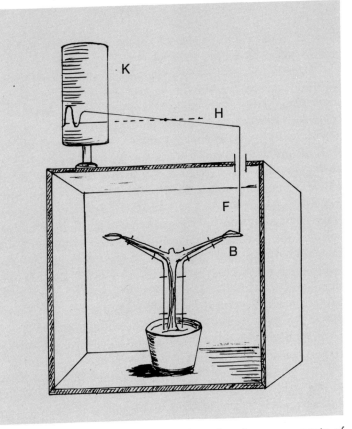

Bünning's early apparatus for recording the sleep movements of bean leaves. Movements of all parts of the plant except the leaf blades were prevented by restraints. A lever (H) was attached to the blade (B) by a thread (F) through a light-lock not indicated in detail. The lever carried a pen which wrote on a rotating kymograph (K). In most uses, the front side of the box was closed so that temperature and light could be controlled.

Bünning and Kurt Stern had worked with very simple apparatus at Frankfurt, and Bünning saw no reason why he could not make do with twenty-five dollars at Jena—as long as he built all of his equipment himself and took complete care of the germination and maintenance of his plants.

In all he spent five years at Jena, during which time he

made two especially important observations about the leaf movements of his scarlet runner beans.

The first of these was that when the plants were kept from the time of germination throughout their lives in constant conditions (either constant light or constant darkness), they completely lacked any regular daily rhythm. If, however, such a plant was given a single stimulus, such as a short light flash for a plant that had been raised in the dark, then it immediately took up a regular daily rhythm that continued for a long period while still in the dark.

The significance of a second discovery will not be fully clear until we come to the Bünning hypothesis itself, but briefly it was that the normal oscillations of the plants brought them into two different kinds of readiness for response to light or darkness. These were qualitatively differing stages in the endogenous rhythm that reversed themselves about every twelve hours. That is to say, that during one twelve-hour period, a plant would respond only to light treatment, and during the immediately following twelve-hour period, it would respond only to a dark treatment.

Also while at Jena, Bünning interrupted his plant studies to pursue a question that Anthonia Kleinhoonte had left unanswered. Recall that she had made no attempt to raise jack beans through successive generations. Indeed, for an organism that takes as long as a plant to go from one generation to the next, such an experiment would have been next to impossible with the facilities available in her time.

There is, however, another organism that is well suited to experiments aimed at learning what happens through many generations. This is the fruit fly *Drosophila*. This fly has many advantages for laboratory culture. It is easy to maintain on a yeasted nutrient medium; it has very large numbers of progeny, so that results of good statistical confidence can be obtained; and its life cycle is a mere eleven days at ordinary room temperature.

For these reasons Thomas Hunt Morgan used it in his critical studies of Mendelian heredity, and in the classical work that he and his students did that led to the gene theory of heredity.[2]

Besides these practical advantages, the fruit fly has a particular characteristic that was ideally suited to Bünning's needs. This characteristic is that the fruit fly has a precise time of eclosion, that is, of the emergence of the adult fly from its pupal case. When it is maintained in normal conditions, the adult fly punctually breaks out of its pupal case very close to dawn. Under laboratory culture the adult flies then mate, the female lays her eggs, the eggs undergo the usual metamorphosis of insects, and eleven days later the adult flies again emerge—just at dawn.

Using *Drosophila*, Bünning could raise a number of succeeding generations in a reasonable time and find out how long a normal eclosion rhythm would persist.

He began by raising several generations of flies under normal alternations of light and darkness at constant temperature, to establish his technique and set up his controls; then he placed his experimental cultures under conditions of constant light and constant temperature. As he expected, the first generation of adults burst from their pupal cases just at dawn.

For how many generations, then, would this continue? Bünning maintained his tedious dawn watches for nearly six months, through fifteen generations. The eclosion rhythm of the fifteenth generation persisted. The adult flies still emerged just at dawn. Throughout fifteen generations deprived of any external clue as to when it was day or when it was night, the last generation was as punctual as the first!

Here seemed proof indeed that the rhythm was internally regulated—that it was truly endogenous.

Throughout his stay at Jena, from 1930 to 1935, Erwin Bünning had been cudgeling his brain for a concept that kept eluding him. At his fingertips were example after example of

2 *The Encyclopedia of Biological Sciences*, Peter Gray, ed. (New York: Reinhold Publishing Corp.; 1961), p. 313.

rhythmic behavior in plants and animals, but he was exasperated because each example seemed to have characteristics peculiar to itself.

True, all of the observations were of rhythmic behavior, but there was no way of looking at each of them as just a representative example of some more general phenomenon. In short, Bünning had no unified hypothesis with which he could tie them all together.

Then on one special day—he does not remember just what day it was—the hypothesis that would relate them all flashed through his mind.

That hypothesis has been called bizarre. It certainly is original, it is highly imaginative, and it offers no simple, familiar physical analogy of which one would say: "Oh, I see! It is like *that!*" There is no other "that" that it is like.

In 1936 Bünning offered the hypothesis for publication, and for several years it attracted very little notice.[3] When biologists did talk about it, they dismissed it as just too fantastic to be worth serious consideration. But at least they did talk about it enough so that for convenience they began to refer to it as the "Bünning hypothesis."

During this same period, other botanists had also been wrestling with the problem of photoperiodism, in an effort to explain why plants behave as Garner and Allard had shown them to do. They wanted to find out which plant organs enabled plants to measure the length of the day, and just how they used them to do it.

Great interest was centered at that time on the biochemical reactions that occur in living materials, and plant hormones and plant pigments occupied the center of the stage. It was not only natural, but really a foregone certainty that someone would suggest that one of the reversible biochemical reactions (a plant pigment, for example, shifting back and forth from one absorbing form to another) was the mechanism

---

[3] Erwin Bünning: "Die endogene Tagesrhythmik als Grundlage der photoperiodischen Reaktion," *Berichte der deutsche botanische Gesellschaft*, Vol. LIV (1936), pp. 590–607.

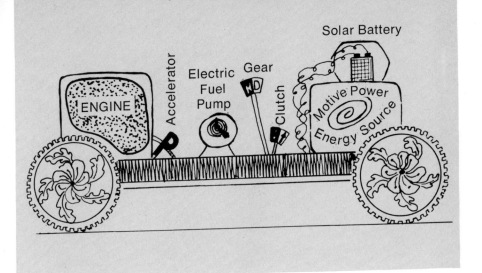

Solar Battery

Accelerator

Electric
Fuel
Pump

Gear

Clutch

Motive Power

Energy Source

ENGINE

In September, 1969, Bruce G. Cumming of the Department of Botany, University of Western Ontario, amused fellow experts at Friday Harbor, Washington, by showing this motive power model analogue of a circadian-photoperiodic system. While this model is intentionally seriocomic, it well illustrates that after nearly forty years, biologists are still in search of a physical model of the original Bünning hypothesis.

that plants used to measure the length of days. By analogy the change of plant pigment from one form to another could be likened to the trickling sand in an hourglass.

To botanists looking for a direct mechanistic answer to the problem, this seemed most attractive. The relative amounts of the two forms of a particular plant pigment called phytochrome could be measured—not only in laboratory glass—but also in living plants themselves. Even though the latter experiments had to be carried out mostly in the dark, and the seedlings had the sickly pale color of a potato sprout grown in such conditions—the etiolated form—the work was vigorously pursued.

And the idea gathered a substantial number of supporters. It was "modern" in that it relied on a biochemical process that could be measured with the latest sophisticated instruments.

It was "comfortable" in that it was based on a well-recognized biochemical reaction.

As it turned out, this hourglass concept ran into some serious difficulties in the end, but in 1936 it had the support of a number of highly regarded authorities, and it took a great deal of courage for Erwin Bünning to propose, in the face of it, a sharply conflicting theory. Yet he felt that he must do so. He might be proved right, or he might be proved wrong, but he could at least hope that his proposal would stimulate research along fresh, new lines.

As it did, for more than three decades.

Now, let us review the range of experimental evidence that Bünning sought to place under a single unified concept:

- Garner and Allard had shown that plants determine the time of year by measuring the length of the light period during a day.

- Kleinhoonte had shown that plant rhythms are very precise; that they continue the same precise behavior when light signals are removed; and that although they can be entrained to abnormal cycles of light and darkness, they return to the normal circadian rhythm as soon as the abnormal stimulus is removed.

- Beling had shown that bees can be trained to visit a feeding station at a regular time of the day, and that they continue to do so even when light, temperature, humidity, and electrical charge in the atmosphere are held constant. She had also shown that bees cannot be trained to a nineteen-hour cycle. Wahl had shown that cosmic rays are not responsible for timing the visits of the bees.

- Bünning and Stern had shown that variations in the electrical charge in the atmosphere are not responsible for the rhythmic movements of plant leaves, but that the leaves can be precisely timed by a red light.

- Bünning had shown that the phase of the rhythm of leaf movements can be shifted by a single exposure to light during a dark period.

- Bünning had shown that the closely timed emergence of fruit flies from their pupal cases persisted through fifteen generations cultured under constant conditions.

A formidable array of observations to be explained by one hypothesis!

Bünning began with two postulates which are perhaps implied in the observations, but which nevertheless need to be stated explicitly. They were:

1. There is a circadian rhythm of some function of plants that is vitally associated with flower induction in both short-day plants and long-day plants.

2. Plants make use of this rhythm for measuring time.

Relating these two postulates with all of the other observations, Bünning arrived at a hypothesis that can be stated in the following way.[4]

The timing mechanism that is responsible for the photoperiodic response of flowering is the same as the mechanism that is responsible for the timing of leaf movements. *This timing mechanism has two alternating phases of about twelve hours each, which may be distinguished as a "light-loving" and a "dark-loving" phase. The light-loving phase is the day phase, and the dark-loving phase is the night phase. Consequently, light falling on a plant during the light-loving phase will enhance flowering, and during the dark-loving phase it will inhibit it.*[5]

Note that the hypothesis itself does not attempt to relate the two phases of the timing mechanism to the behavior of bees or fruit flies, but the belief that such timing mechanisms are also entirely endogenous is closely associated with it.

It is therefore understandable that Bünning should hold the strong belief that the timing mechanisms of most, if not all organisms, are almost certainly internal. That is why he has sometimes shown a choleric impatience with those who argue that the overriding time-giver for rhythms originates in forces outside the organism.

[4] Bünning himself has no preferred wording for the hypothesis. Erwin Bünning to Ritchie R. Ward, personal communication, October 28, 1969.
[5] Beatrice M. Sweeney: *Rhythmic Phenomena in Plants* (New York: Academic Press, Inc.; 1969), p. 76.

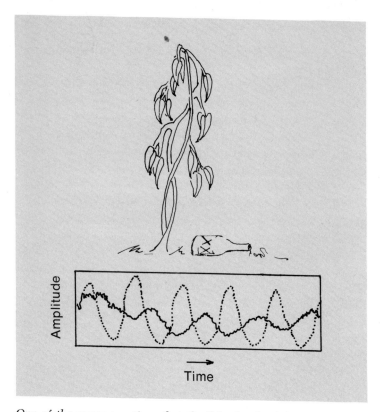

One of the many questions that the Bünning hypothesis raised was how plants would behave under the influence of drugs. In 1960 S. Keller measured the diurnal leaf movements of beans under the influence of alcohol, in continuous dim light. In the graph, the dotted line shows the regular normal behavior of a sober bean; the much slower and erratic solid line shows the influence of alcohol. This offered further proof that there is a circadian rhythm, and that it is markedly disturbed by alcohol.

With the publication of his hypothesis paper, Bünning's career at Jena drew to an unhappy close. A world depression had brought crisis to Germany. Loans were canceled and foreign investments withdrawn. Banks closed their doors across the land and riots shook the cities. As Hitler's power mounted, the school system was integrated with the German Youth Movement. German universities, traditionally renowned

Erwin Bünning in one of his greenhouses at the Biological Institute, University of Tübingen, Germany. The kalanchoe plants are here being illuminated by a bank of fluorescent lamps.

for their academic freedom, turned into agencies for propagating the racial myths of Nazism. Increasing pressure from Nazi students finally drove Bünning from Jena.

Before long he took a post at the University of Königsberg (now Kaliningrad); but when World War II ended, and he was offered an appointment at the University of Tübingen, he gladly accepted. He has remained there since 1945, continuing his researches on endogenous rhythms in the Institute for Biology there.

Today, thirty-five years after its first publication, the Bünning hypothesis has still not gained full acceptance. Some biologists continue to think that it is so bizarre that they would rather not commit themselves. Some have found their own efforts to offer a better idea so frustrating that they accept it with some reservations. Some have found that it suggested rewarding new experimental approaches. For example, Karl C. Hamner has said: "I believe my own work and the work of my students provides fairly conclusive evidence that endogenous rhythms participate in the photoperiodic reaction, and that something very close to Bünning's hypothesis may be the case."[6]

While biologists are thus divided on accepting the Bünning hypothesis in toto, they fully agree (with some violent exceptions) with the two postulates that we noted earlier were implied in it.

These are so important that they bear repeating.

First, plants (and by extension, probably animals also) have endogenous circadian rhythms. This means that they have rhythms that repeat over and over at approximately twenty-four-hour intervals; and most important, that the rhythms are self-sustaining. Organisms have the ability, that is, to *draw energy from a constant source and convert it into more useful alternating cycles,* which are then displayed as rhythmic phases of activity like the hands of a clock returning to midnight once each day.

[6] Karl C. Hamner: "Photoperiodism and Circadian Rhythms," in *Cold Spring Harbor Symposia on Quantitative Biology,* Vol. XXV (1960), pp. 269–77.

Second, organisms use these endogenous rhythms to measure time.

These were really the concepts that stirred all the furor. The idea that a plant could possess a clock was just too fantastic! Too fantastic, that is, until another fifteen years had passed, and evidence began piling up that birds and bees navigate by the sun and the stars, and that this navigation would only be possible if they too possessed internal clocks. As that evidence mounted, a great upsurge of interest in biological clocks ensued, and new research was undertaken on organisms ranging from one-celled plants to man.

How will the hypothesis finally fare? Bünning himself takes a long view with amused tolerance. "Until about 1939, it was not even worthwhile," he says, "to discuss with others a nonsense like that. Endogenous rhythms were looked on as mystiques, and their combination with photoperiodism as metaphysics. In the decade from 1939 to 1949 it was worthwhile to discuss these ideas—just to raise the question: is it after all *complete* nonsense? In the 20 years since 1950 more and more people found: there is something in it."[7]

There is indeed something in it.

Throughout his years at Tübingen Bünning has continued his researches, broadening and deepening biology's understanding of photoperiodism. He has long since attained the stature of a world authority, and as the acknowledged father of the clock concept he is invited to tell of his latest thinking at every conference and symposium in the field. At the Cold Spring Harbor Symposium on Biological Clocks in 1960 he gave the opening keynote address, and at the 1969 International Symposium on Biochronometry sponsored by the National Academy of Sciences-National Research Council and N.A.S.A., he presented a paper entitled "The Importance of Circadian Leaf Movements for the Precision of Photoperiodic Time Measurement."[8] The second German edition

[7] Erwin Bünning to Ritchie R. Ward, personal communication, October 28, 1969.
[8] In press.

In recent work Erwin Bünning studied the sleep-wake rhythms of cut blossoms of the brilliant scarlet kalanchoe under constant conditions. ABOVE, the blossoms open when it is day outside, and close when it is night, BELOW. Bünning's present apparatus is a far cry from the austerely simple device that he used at the University of Jena. Now photoelectric cells actuate a continuous electronic recording instrument.

of his definitive monograph on the circadian clock was published in 1963 and translated into English a year later.[9]

Whether he suspects it or not, there is a growing feeling among biologists that if anyone in the rhythm field should ever receive a Nobel prize, that prize should go to Erwin Bünning.

[9] Erwin Bünning: *The Physiological Clock*, trans. by Lars Lörcher and Frank Salisbury (New York: Academic Press, Inc.; 1964).

Probably the knottiest
problem in all ornithology
is how a bird finds
its way home.
        JOEL CARL WELTY,
              1963

9. Kramer: Solar Navigation by Birds

Once in a very long time in the history of science, a researcher looking for one answer finds an entirely different and far more important one. This has been called serendipity after Horace Walpole's three princes of Serendip, who were always making delightful discoveries by sheer accident.

Far more rarely, a scientist finds a brilliant answer to the very problem that he set out to solve, and then further finds to his delight that that answer holds implications of deeper significance than the original problem seemed to suggest. That might be called serendipity in spades.

In just such a way Gustav Kramer made a discovery so provocative that many biologists in world centers of research began to lay aside their ongoing work to join the burgeoning ranks of those who were challenging the riddle of living clocks.

Kramer was born in Mannheim, Germany, in 1910 and received his education in biology at the Universities of Freiburg and Berlin. His first research interests were in the physiology of lower vertebrate animals, and his work in that field showed such promise that at the age of twenty-seven he was appointed acting head of the department of physiology at the Zoological Station at Naples.

World War II then intervened, but when it ended Kramer was called to lecture on physiology at the University of Heidelberg. There, in addition, he began his studies of how birds orient themselves in flight.

Again his research began to attract attention, and after two years at Heidelberg, Kramer joined the staff of the Max-Planck-Institute for Marine Biology at Wilhelmshaven. There he pursued his work for some ten years and built up a school for orientation studies that carried on work in the United States as well as in Europe.

While Kramer was a remarkably able physiologist, he was also a keen field naturalist, fascinated by the many avenues along which the behavior of birds could be studied. Among these, his own driving interest continued in finding out how birds determine their courses during long flights; but he recognized orientation as just one problem among many in the mystery of animal migration.

The Institute for Marine Biology at Wilhelmshaven lies on the western shore of Jade Bay, looking out over the cold North Sea. On long walks along the water's edge, Gustav Kramer watched the swift homing flights of the sea birds, and continued to ponder the age-old problem of migration, and the wonder of the accuracy of migrating birds in finding their goals.

He marveled at the feats of the arctic tern, a phenomenal flier that nests within a hundred miles of the North Pole, and then in the fall flies over Canada, across the Atlantic, down the west coast of Africa, and around the Cape of Good Hope to winter feeding grounds below Port Elizabeth.

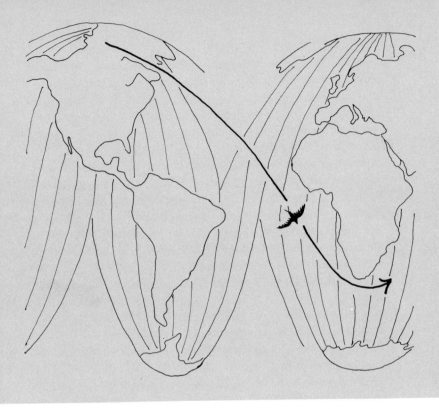

The extraordinary migratory flight of the arctic tern. After nesting near the North Pole it flies south over arctic wastes, across the Atlantic, down the west coast of Africa, and around the Cape of Good Hope to winter feeding grounds.

Nor is the tern alone in its navigational skill. The New Zealand bronzed cuckoo flies 1,200 miles across the Tasman Sea to Australia, and from there 1,000 miles north over the Coral Sea to tiny wintering areas in the Bismarck Archipelago and the Solomons. Even more astonishing is the fact that the cuckoo young, never having flown the course, do so alone; their parents precede them by at least a month.

Banded white-crowned sparrows return to the same bush in the garden of Professor L. Richard Mehwald in San Jose, California, after having flown 2,200 miles from their summer breeding grounds in Alaska.

The mystery of such pinpoint flight accuracy has intrigued

biologists for a very long time, and over the years they have offered a variety of explanations for it. A few let their imaginations get quite out of hand—and small wonder. The problem was extraordinarily complex, and there was as yet no way to put it to scientific test; no controlled laboratory experiments had even been thought of.

In fact, actual observations of bird flight beyond the normal range of vision were not begun until 1940, when Donald R. Griffin learned to fly a light plane, and began following the homing flights of herring gulls and gannets.[1]

When, therefore, Gustav Kramer revealed to an international congress of ornithologists the results of his laboratory experiments on bird orientation, the congress was amazed and delighted. Roger Tory Peterson exclaimed: "How thrilling it was in 1950 to hear Gustav Kramer tell of his controlled experiments with starlings, indicating that the ultimate source of their orientation was the sun."[2]

Kramer had found a brilliant answer to a particularly thorny problem in a field hedged about with difficulties. That field—the whole field of animal migration—is very broad, and direction finding is, of course, only one aspect of it. But unusual insight into one such aspect most often leads to clarification of the whole field. As an impressive world ecological phenomenon, migration has been put in excellent perspective by Joel Carl Welty: "Animals," he wrote, "unlike green plants, cannot root themselves in one place and make their own food. They must move about and search for it, and as soon as they have exhausted the food in one place they must move on to another. With their unparalleled mobility, birds have explored almost the whole earth, searching not only for food, but also for territory, mating sites, and other necessities for survival."[3]

---

[1] Donald R. Griffin: *Bird Migration* (Garden City, N.Y.: Doubleday and Co. Inc.: Natural History Press; 1964), p. 100.
[2] Roger Tory Peterson: "Forward" in Jean Dorst: *The Migrations of Birds* (Boston: Houghton Mifflin Company; 1962), p. viii.
[3] Joel Carl Welty: *The Life of Birds* (New York: Alfred A. Knopf, Inc.; 1963), p. 445.

When animals migrate from one place to another they secure a number of major advantages, both for themselves as individuals, and also for the preservation of their respective species.

Obviously when birds summer in the north and winter in the south, they are arranging for a favorable climate. At the same time they obtain a variety of diet that would not be available in one place. Exposure to disease is reduced upon abandoning soil that may have become infected and thus a source of contagion. Harsh struggles with the environment for survival—that is, ecological extremes—are reduced. And most important for the species, migration promotes rigorous natural selection: migratory animals develop a greater range of adaptability than sedentary ones, and dispersal promotes crossbreeding on the one hand, while on the other it tends to isolate small populations in which the rate of evolution is speeded.

But these migrations, as we have seen, are directed toward specific goals—vanishingly small on the surface of the earth, and as far as half the world away. Such accuracy would be inconceivable without presuming some guidance system analogous to that by which a torpedo seeks its target. To presume such a situation, however, offers no particular difficulty. Maxwell Maltz believes that all living organisms possess an extraordinary mechanism—like the servomechanism of the torpedo—that directs them toward desirable goals: "Every living thing has a built-in guidance system or goal-striving device," he wrote, ". . . to help it achieve its goal—which is, in broad terms—to 'live.' In the simpler forms of life the goal 'to live' simply means physical survival for both the individual and the species. The built-in mechanism in animals is limited to finding food and shelter, avoiding or overcoming enemies and hazards, and procreation to insure the survival of the species."[4]

It is extremely important, however, to recognize that no

[4] Maxwell Maltz: *Psycho-Cybernetics* (Englewood Cliffs, N.J.: Prentice-Hall, Inc.; 1960), p. 14.

such servomechanism can function without a steady flow of information from the outside world. The homing torpedo must receive signals reflected back from its target or it will miss. And animals must receive signals of some sort from the environment, or their goal-striving devices will fail.

What kinds of information can the environment provide?

For the orientation of bird flight, biologists have given serious consideration to two kinds: information accessible to the known senses, and information that could be detected only by senses that are not known.

But no matter how the migrating bird receives the information, that information must be such that the bird can solve three problems.

First, where it is and in which direction it should start from there.

Second, how to maintain its direction as it flies, or to modify it if necessary to arrive at its destination.

Third, how to recognize the destination when it gets there.

Is there any single sense, known or unknown, that might answer all of these questions? Let us consider the possibilities.

Every object on the surface of the earth emits thermal radiation. Hot objects emit radiation of high intensity and short wave length, and cool objects emit lower intensities at longer wave lengths. Therefore both the frequency and the intensity of radiation at the pole will be far different from those near the equator. It is obvious to suggest that long distance migrators might be able to sense this difference. As Donald R. Griffin has remarked, this would be a delightfully simple explanation of bird orientation.[5]

However, three compelling facts stand in the way. Radiation travels in straight lines rather than following the curvature of the earth; therefore radiation from even 100 miles away would arrive at a point far above a bird's normal flight altitude. Also, thermal radiations are seriously distorted by

[5] Griffin: *Bird Migration,* p. 77.

large land features such as lakes, deserts, forests, and cities. These distort the original radiation by adding "noise," as information specialists call it. Finally, no one has ever been able to show that birds can sense differences in thermal radiation.

But these are ordinary heat waves. How about something more esoteric? The earth's magnetic field, for example. That too has been proposed as a bird compass. Theoretically the idea has considerable appeal. We have parallels of magnetic field strength that run roughly with the geographic parallels, and if a bird could actually sense differences in magnetic field strength, it could determine its latitude from that. Another suggestion depends on the declination of the earth's magnetic field; if a bird could detect this, its compass needle would be horizontal at the equator, and nearly vertical at the pole. This also might tell the bird where it was. But again the facts stand in the way. Experimentally, birds show no spontaneous reaction to magnetic fields even much greater than the earth's; moreover, researchers have never been able to train birds to respond to magnetic fields.

What other features of the earth's environment might one consider? The earth's rotation is an obvious one. The angular velocity of rotation is such that a point on the surface of the earth (if it is near the equator) has a speed of some 1,000 miles an hour. If a bird flies east at 60 miles an hour, its true speed will then be 1,060 miles an hour; and if it flies west its true speed will be 940 miles an hour. The bird therefore would be able to sense its direction of flight and its latitude from differences in the surrounding physical forces; and at least one physicist has suggested that birds can detect such small differences through mechanical pressure of the air against their wings.

But what if the bird is not flying? In one case that resulted from an accidental escape, geese pinioned so that they could not fly traveled for many miles by land in the same direction they would have taken in migratory flight. Moreover, it has been convincingly shown, as we shall soon see, that caged birds can sense direction very well indeed.

Despite all such evidence, however, it is still not possible to conclude that no sense, as yet unrecognized, participates in bird orientation. Michael Menaker of the University of Texas has shown that blinded birds can perceive light in some non-visual way.[6]

It may well be that the ever-increasing sophistication of research concepts and research techniques will someday reveal some hitherto unsuspected sense that will throw new light on the whole problem. But that time is not at hand. Meanwhile, the most promising advances have been made on the simple assumption that birds depend primarily on their acute sense of sight.

Nevertheless, the range of possibilities that have been seriously considered gives some impression of the complexity and difficulty of the problem that Gustav Kramer faced in 1949.

Until then, the main shortcoming in experimental work on long distance orientation had been the long distance itself. A bird's spontaneous choice of flight direction could be observed only in the field. New advances awaited the birth of a new experimental technique.

Kramer turned the problem over and over in his mind. Gradually a new idea began to take form. It was a vague idea to be sure, and it might lead only into a blind alley, but he thought it would be worth following, and he set to work.

It had long been known that during migratory seasons, caged birds exhibit what are called "flight fidgets." They flutter about restlessly, and as they do so, they point repeatedly in a particular direction. Could that be the direction they would take if they were free to fly? That question Gustav Kramer set out to answer.

---

[6] Michael Menaker: "Extraretinal Light Perception in Birds," *Proceedings of the Friday Harbor Biochronometry Symposium* (September, 1969), in press. When he presented this paper Menaker humorously said that his preparation of a bird for these experiments required an operation of the utmost precision and delicacy: he plucked three feathers from its head.

Among the birds that commonly do well in cages, the European starling seemed a good choice. Young birds were easy to collect, and Kramer found that when he raised them by hand, they became very tame and easy to train.

Soon the Wilhelmshaven laboratory was stocked with the yellow-billed young birds, whistling musically or rasping harshly by turns; and Kramer waited impatiently throughout the summer for the fall migratory season to begin.

As the crisp October days approached, he kept a constant watch over his starlings throughout the daylight hours. (He did not need to watch them during the night, for starlings are diurnal migrators.) From Wilhelmshaven, the normal fall migration direction of starlings would be to the southwest. Would the caged starlings show any significant preference for this direction when they began to get the flight fidgets in their cages?

Kramer was not long in finding out. In October his captive birds, with all their nervous fluttering, tended to remain in the southwest corner of their cages, and to face to the southwest direction.

What clues or points of reference were the birds using? Did they relate themselves to some physical feature, such as a hill or a tree? Kramer tested this by moving the cages to other areas with different landmarks. He also enclosed the cages part way up, so that a starling could see nothing but the sky. His birds continued to favor the southwest direction as strongly as before.

The next spring, when the migratory direction would be to the northwest (this is not, for starlings, the exact opposite of the direction of fall migration), the caged birds pointed to the northwest.

Here, then, was the exciting basis for the experimental method that Kramer had long sought. His next step was to design and build equipment with which he could make quantitative measurements—thousands of observations to ensure statistical significance for his results. If he foresaw the tedious, body-and-brain-wearying demands that lay before him, he brushed them aside.

His first piece of apparatus was a circular cage, so designed that a bird inside it had no way of knowing which way it faced; the inner surface of the cage was perfectly symmetrical. From a central perch, a bird with the flight fidgets would fly in one direction to the edge of the cage and back to the perch again. The bottom of the cage was of clear plastic, so that Kramer could lie beneath it and observe the bird's position. And as a reference, he marked off the transparent plastic into a number of sectors so that he could record the bird's exact position at any instant.

The most important experimental variable that he wanted to test was the direction from which the light fell on the cage; but he also wanted to try varying the intensity of the light. To provide for both conditions, he built a six-sided pavilion in which to hang the observation cages. Each of the six sides of the pavilion had a window provided with a shutter; each window was also provided with a mirror so that light could be reflected into the cage at right angles to the normal direction. Finally, Kramer provided for rotating both the cage itself and an opaque screen that could be placed around the pavilion.

When all was in readiness he stretched out directly beneath the transparent cage bottom, notebook and pencil in hand, and punctually *every ten seconds* he wrote down which of the marked-off sectors the bird occupied. This trying routine he continued for at least an hour every morning, and after many such mornings he was quite sure that neither the equipment nor his own presence beneath it upset the starlings in any way.

Just as his starlings had previously done in their cages in the open, the birds in the pavilion now pointed in the spring migration direction—the northwest. That is, they pointed that way when the windows of the pavilion were open on all sides.

The great importance of that advance will now be clear. No longer need researchers be handicapped by the uncertainties and inaccuracies of field observations. Those could thankfully be left behind. The direction of free migratory flight could now be determined under laboratory conditions that allowed the experimenter to vary his controlled conditions in any way that he chose.

The question that Kramer was most curious about just then was how the birds would respond if, instead of entering directly, the light were reflected at right angles by mirrors. In that situation, the position of the sun would seem to a bird in the cage to have been rotated by 90°.

What difference would that make to the bird?

Once more Kramer crawled under the observation cage and again began doggedly noting more hundreds of observations. "First 10 seconds, bird in sector 8; second 10 seconds, bird in sector 9; third 10 seconds, bird in sector 7; fourth 10 seconds, bird in sector 9; fifth 10 seconds, bird in sector 8, . . ." and so on and on and on, until he had made 3,500 and more recordings in the course of an hour. Soon Kramer began to feel confident of his results. But would the skeptical world of science accept them? He thought very decidedly not, because the outcome was so startling. So he took up his tedious task of observing and recording once again, for he wanted unassailable proof.

When he did announce his findings, the world of science was indeed astonished, but it could hardly be skeptical in the face of Kramer's mass of data.

What amazed scientists was that the starlings changed their direction by 90° when the light from the sky was changed by 90° by mirrors. The birds were not governed by any "natural inherent tendency" to point in the direction of their normal migration. *They did not know what that direction was unless they could take their bearings from the sun!*

If they could see the sun directly, they assumed the normal direction. If the sun seemed to them to be somewhere else, then they changed direction to keep their bearings on it.

In that first experiment, Kramer had set the mirrors so that the birds turned 90° in the clockwise direction. Now he

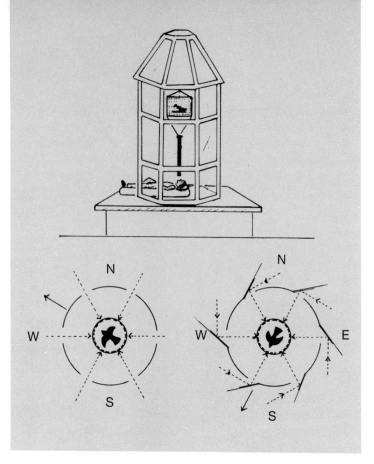

ABOVE, *the apparatus of Gustav Kramer that made possible the first laboratory observations of a bird's choice of migratory direction. Kramer observed the bird's position through the transparent plastic bottom of the cage. When the starling could see the real position of the sun, it pointed to the normal northwest migratory direction,* BELOW LEFT. *When the sun's rays were deflected through 90° by mirrors, the bird changed its direction by the same 90°,* BELOW RIGHT.

removed the mirrors, and the birds returned to their normal flight direction. Finally, he reversed the mirrors, and the birds dutifully turned 90° counterclockwise.

But that was a mere beginning.

Scientific discoveries have a way of answering one question and forthwith asking many more. Kramer sought to answer some of the new ones by changing his experimental conditions in various ways. He rotated opaque screens around his pavilion, so that the birds could see most of the sky, but nothing else. He rotated the cage itself. And he enclosed the pavilion in screens that let various amounts of light through, to imitate cloud covers both light and heavy.

No matter how he changed the conditions, his starlings in the spring always pointed in the direction that they took to be northwest, whenever they had access to an apparent position of the sun.

Now Kramer paused to review his results. They seemed positive and convincing, but they might be questioned on one score: the periods of observation were not very long—only about an hour each morning. They were nerve-wracking in their demands for continuous attention. They were also limited by the difficulty of obtaining starlings that would co-operate as well as those in the experiments that he had just completed. Not all of his birds were willing to make a spon-taneous choice of migratory direction.

Kramer mulled over ways that might avoid these dif-ficulties. He was, of course, familiar with Beling's earlier work that had shown that bees could be trained to seek food in a particular direction. Could birds be trained in the same way? If they could, then it would no longer be necessary to depend on a bird's willingness to make a spontaneous choice. If a bird would learn the direction in which food lay, then Kramer could vary his experimental conditions at will, and find out what effects the new conditions would have on the bird's behavior.

To test the promise of this, he built a circular training cage that was again perfectly uniform viewed from the inside. Now, however, he added twelve feeding stations equally spaced around the outer edge. They too all looked the same, and Kramer masked them with slitted rubber membranes so that a bird could not know which one contained grain until he thrust his bill through the slit and found out directly.

Now Kramer began the first phase of a crucial experiment by training a bird to seek food at the east side of the cage.

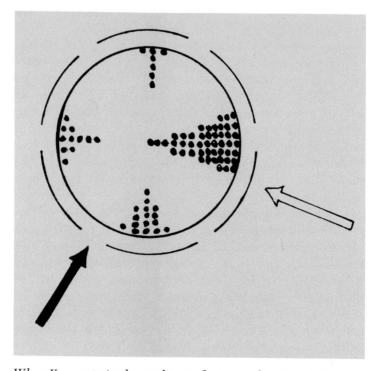

When Kramer trained a starling to fly in one direction at the same time every day, for example when the sun shone from the direction of the hollow arrow, the bird then knew which way to fly at any other time of day, for example when the sun shone from the direction of the solid arrow. The dots indicate individual observations of the bird's position.

At seven in the morning he put grain in the east feeder and none in any of the others. The bird showed great persistence, and after a series of random trials, always finding food in the east feeder but never in the others, it began to learn that food lay to the east and nowhere else.

After 28 days of training between 7 A.M. and 8 A.M. the starling had learned its lesson very well. With only an occasional error—and that error in the right general direction—it sought its reward due east.

Now came the critical test.

After the 28-day training period, Kramer moved the

orientation cage 7 miles away, and *at 5:45 in the afternoon* he again put grain in the east feeder.

How would the bird respond?

During the entire period of training in the morning, the sun had been slightly to the right of the east feeder. Now, late in the afternoon, it was behind the west feeder. Would the bird now think that food lay in the direction of the sun? Would it head west for food? Tensely Kramer waited. The starling fluttered briefly, appearing undecided, and then with only an occasional error, it headed for the east feeder and its reward.

Somehow the bird knew that even though it should bear toward the sun to go east in the morning, it should now place the sun directly behind it to go east in the afternoon!

How completely unexpected this was has been highlighted by one of the top authorities in the field, Donald R. Griffin. "If anyone had discussed this experiment with biologists or other scientists in advance of Kramer's actually carrying it out, it would have been agreed rather generally that a bird could be expected to find the correct feeder, provided the feeder remained in the direction of the sun or at some constant angle to it.[7] It would have been difficult to believe, however, that a bird could easily and rapidly compensate for the sun's movement through the sky."[8]

To buttress his conclusions still further, Kramer designed an experiment of extraordinary brilliance. First, under the open sky, he trained a starling to seek food in a feeder to the west of its cage no matter what the time of day. He then placed a protective tent around the bird's cage so that light from the real sun was excluded, and outside the protective tent he placed an artificial sun *so that it shone always from a fixed position in the west.*

Confronted with a "sun" that shone continuously from a fixed position, what would the poor bird do now? As Kramer watched, bursting with anticipation, the starling calmly as-

[7] One of Kramer's collaborators, Ursula v. St. Paul, later trained one starling to head into the sun, but it took "Heliotrope" seven weeks to learn that lesson.

[8] Griffin: *Bird Migration*, p. 125.

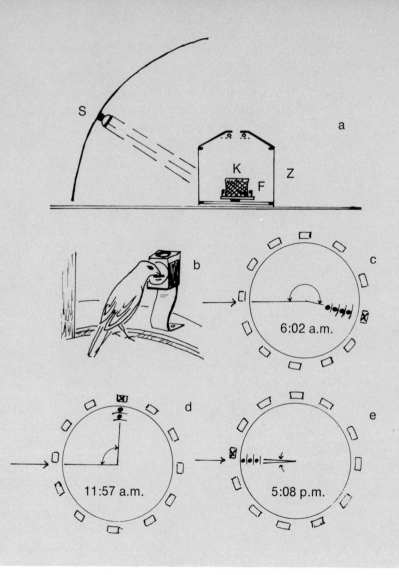

Kramer's apparatus (a) for studying a starling's choice of direction under a fixed sun (S) included the cage (K), feeders (F), and protective tent (Z). The starling was first well trained under the open sky to seek food (b) in the feeder at the west side of the cage. The protective tent (Z) was then put up to exclude light from the real sun, and the fixed artificial "sun" (S) turned on. The bird assumed that (S) was moving as a sun should, and sought food in the east in the morning (c), to the north at noon (d), and to the west late in the afternoon (e).

sumed that it was simply seeing the real sun, and proceeded to act as though that "sun" were moving as it ought to move through the heavens. Having been trained to seek food in the west feeder at any time of a normal day, *it now sought food in the east feeder at 6* A.M., *in the north feeder at noon, and in the west feeder at 5* P.M.*!*

Could anyone now doubt that this darkly iridescent bird could count off the time of day to the minute?

These, then, were the amazing discoveries about bird orientation that Gustav Kramer reported in the early 1950's. And although they very quickly brought him world recognition, he looked at these accomplishments through the eyes of an uncommitted and cautious man. He felt that a very great deal more work needed to be done before a true understanding of bird orientation could be achieved.

Since he had now shown that birds determined direction by referring to the sun and compensating for the sun's movement, he could say that the bird very literally possesses a sun-compass. And that it uses it in exactly the same way as a navigator uses a magnetic compass to set his course. But that was only part of the problem of reaching a definite goal. For a man's skills in navigation to be of any use to him, he must also have a map, and he must know where he is on that map. So too must a bird possess a map to find its migratory goals as accurately as it does. And practically nothing was known about the map.

What experiments, then, could Kramer devise to learn something about the map and how the bird uses it? Again he turned to the literature. Prominent among other bird navigation studies that were being carried on at the time was the work of G. V. T. Matthews, who had made extensive studies on the homing behavior of pigeons in England. Matthews had written the definitive monograph on bird navigation and Kramer turned to Matthews's work to see what ideas it might offer.[9]

[9] A completely revised and updated second edition of this monograph is now available—G. V. T. Matthews: *Bird Navigation* (Cambridge: Cambridge University Press; 1968).

Gustav Kramer releases homing pigeons from the tower of old Gleiberg Castle, near Giessen.

Very soon he found himself in sharp disagreement with Matthews's views on exactly how a bird uses the sun to navigate, but that was far less important than the promise of the techniques that Matthews had devised.

Matthews was learning a great deal by taking homing pigeons from their nesting site to carefully chosen release points, and then observing their flight through field glasses for the few minutes it took for the dwindling image to disappear. This was then carefully compared with the bird's arrival at the nesting site.

Kramer, in his usual thoroughgoing fashion, laid out an ambitious program of experiments of his own—a program which he began, but which, tragically, he was not able to finish. He began by making many observations of the flight of homing pigeons that he released from the tower of old Gleiberg Castle, near Giessen. With Ursula von St. Paul at Wilhelmshaven, and J. G. Pratt of Duke University, he did two-directional experiments in North Carolina under the joint sponsorship of the U.S. Office of Naval Research and the Duke Laboratory of Parapsychology. Hoping to develop birds that would perform more expertly than those he had used so far, he began collecting wild rock pigeons in mountainous Calabria, at the southern tip of Italy. While climbing there on April 4, 1959, he fell to his death.

Science lost an outstanding researcher just as his greatest achievements lay at hand.

W. H. Thorpe, a noted authority on animal learning and instinct, wrote the obituary for the British Ornithologists' Union. In part he said:

> He was inclined to believe that a long further period of collecting facts without being too much tied to any one theory was necessary before substantial progress could be expected.

> Not only was Gustav Kramer an able physiologist, he was most conspicuously a first-class field ornithologist with an appreciation of the fascination of many different branches of the study of birds. It was in this combination of field naturalist and physiologist that his great strength lay.

> In 1958 he was appointed head of a new section of the *Max-Planck-Institut für Verhaltensphysiologie* at Tübingen, and in 1959 became in addition director of the *Vogelwarte Radolfzell* (formerly *Rossiten*). It is hard to see an adequate substitute for him in either capacity.

> Personally he was a man of great charm, with a somewhat impish sense of humor, and an intolerance of all stuffiness and formality. He was an outstanding athlete and a keen swimmer and would take standing jumps and dives which were enough to cause palpitations in those who watched. He was also amazingly hardy: when staying in Oxford for the Ethological Conference in 1955 he arrived back late one night to find the Holywell Hostel of Balliol College already bolted and barred. He was loath to wake the custodian and diffident about climb-

ing in so he calmly spent a damp September night sleeping on a flat tombstone in the neighboring churchyard of St. Cross.[1]

But before his untimely death, Gustav Kramer had made a revolutionary change in biological thinking. Let us see more precisely just how.

Recall that Kramer established beyond question that birds orient themselves by the sun. Recall too that they readily compensate for the sun's movement across the sky. How could they accomplish this compensation? The only conceivable way would be through the use of an internal clock. A clock so accurate, indeed, that it could be compared only to the chronometer on which a human navigator depends.

Here was proof of a living clock that left no possibility of evasion. It was no longer possible to say, as it was with Garner and Allard's discovery of photoperiodism: "I grant that a plant can measure the length of a day, but I do not need to concede that it has a clock. A plant could measure time in other and more reasonable ways." It was no longer possible to say, as it was with the Bünning hypothesis: "I am willing to grant that Bünning's findings are valid, but I can explain them just as well by a simple biochemical reaction. I do not need Bünning's endogenous rhythms to explain Bünning's results."

From Kramer's demonstration of a living clock there was no such escape.

Few biologists wanted to try. Most were both convinced and fascinated by the new avenues for research that the concept opened up. A great upsurge of research along unprecedented lines began in what is now referred to as the science of biorhythms.

In short, a new science was born.

In contemplating his achievement, Kramer's friends[2] turned to the words of Goethe:

[1] W. H. Thorpe: "Obituary, Gustav Kramer," *Ibis*, Vol. CI (July–October 1959), p. 510.
[2] Erwin Stresemann: "Gustav Kramer," *Zeitschrift für Tierpsychologie*, Vol. XVI (1959), p. 257.

Was kann der Mensch im Leben mehr gewinnen,
Als dass sich Gott-Natur ihm offenbare,
Wie sie das Feste lässt zu Geist verrinnen,
Wie sie das Geisterzeugte fest bewahre.[3]

[3] The poem can be roughly translated as: "What more can man profit from life than to apprehend how God reveals himself through nature, and how constant nature is, imbued with the spirit." *Goethes Sämtliche Werke, Jubiläums-Ausgabe* (Stuttgart and Berlin: J. G. Cotta'sche Buchhandlung Nachfolger; 1902–7), Vol. I, p. 286.

10. Von Frisch and Renner: The Clock of the Bees

Karl von Frisch, the venerated world authority on the social life of bees, received unprecedented acclaim from the world of science for an experiment that he did not want to do.

The acclaim, accorded von Frisch throughout the world of biology and far beyond, has been memorably expressed by Archie Carr:[1]

> One of the dramatic zoological events of this century was the discovery by Karl von Frisch of Germany, of the sun compass

---

[1] Carr is Graduate Research Professor of Biology at the University of Florida. An expert on animal navigation, he has fastened helium-filled balloons to the backs of green sea turtles to chart their migrations from Key West to Mexico. He believes that the turtles may navigate by the sun.

in honey bees. For ages men have wondered how bees could fly so straightly to a source of nectar and back to the hive again. The brilliant work of von Frisch revealed that bees hold to their beelines by taking bearings on the sun, and take its apparent motion into account in plotting their beelines at different times of the day. He capped it all in a sensational way by discovering that a bee who has found a good patch of blossoms can tell her colleagues where it is by means of a little dance she does. In the dark, on the vertical wall inside the hive, she makes and reverses a small upright circle. Every so often she dances a diameter across it; then she circles some more. The angle the diameter makes with the vertical tells other bees which way to fly. The vertical is the sun's azimuth position—its compass bearing from the hive—and the danced diameter shows the angle that the flight path to the food makes with the sun's direction. All another bee has to do is mentally to lay the dance pattern out flat, with what was the vertical now pointing in the direction of the sun, and then fly away in the direction the diameter shows. Of course, as the day wears on, the angle between the beeline and the sun's bearing changes. The bee dancing the pattern knows this and keeps narrowing the diameter angle, if it is morning, or widening it if it is afternoon.

This is called the *sun compass sense*. Discovering it was one of the most important advances ever made in the study of animal orientation. That the discovery should have included finding bees communicating with each other in such an elaborate way is an almost incredible coincidence.[2]

Von Frisch's work is thus seen to strike an astonishing parallel to Gustav Kramer's discovery of the sun compass in

---

[2] Archie Carr: *Guideposts of Animal Navigation, Biological Sciences Curriculum Study Pamphlet 1* (Boston: D. C. Heath and Company; 1962), p. 25. Used by permission of the Educational Programs Improvement Corporation, Boulder, Colorado. Although men have wondered how a beeline can be so straight, they also made practical use of it long ago. Karl C. Hamner has reported the frontier lore that the American pioneers located bee trees by setting out a small dish of honey and watching carefully as the honey-laden bees left the dish. The direction of departure provided a line to the bee tree. ("Experimental Evidence for the Biological Clocks," *The Voices of Time*, J. T. Fraser, ed. [New York: George Braziller, Inc.; 1966], p. 650.)

birds. Even more remarkable is the fact that neither Kramer nor von Frisch had the slightest idea of what the other was doing. Nor did they find out until the work of each was done and published.[3] This in spite of the fact that they began their work at almost exactly the same time—the fall of 1949—and that both were working in Germany in research institutes little more than four hundred miles apart.

Still, it is not hard to see why the two scientists should have gone their independent ways. While both were biologists, their specialties were entirely different: Kramer was a bird man, and von Frisch was a bee man; and each had his hands more than full with his own problems.

Then too, the particular problems that they had set out to solve were very different. Kramer wanted to explain the basis of bird navigation, and in the course of doing so, he discovered the sun-compass. From this came the mandatory conclusion that birds can tell time. Von Frisch, on the other hand, had been convinced for many years that bees can tell time, and in trying to find out how they did so, he was driven to the conclusion that bees have a sun-compass. Thus both scientists arrived at the same concept from opposite directions.

Von Frisch's conclusion came, strangely, from an experiment that he did not want to do, and that he kept putting off time after time. Let us see how this came about.

Recall that from his early years at Munich von Frisch had been fascinated by the nineteenth-century observations of Forel in Switzerland and von Buttel-Reepen in Germany that bees came for sweets only at such times as they were used to finding them. Recall too, that von Frisch's graduate student, Ingeborg Beling, had shown in experiments of unusual precision not only that bees are remarkably accurate time-keepers

[3] Karl von Frisch: *The Dance Language and Orientation of Bees* (Cambridge: Harvard University Press; 1967), p. 338. There von Frisch wrote: "Our experiments were conducted simultaneously, but without our knowing of one another. Afterward each of us found pleasure in the unexpected discoveries of the other."

under normal conditions, but more important, that their time-keeping ability continues under constant conditions of light, temperature, humidity, and atmospheric electrical charge.

This being the case, only two explanations of time-keeping seemed possible. Either the clock of the bees was entirely internal, and required no clue from outside, or it was governed by some unknown force in the environment.

Such a force would have to have a regular daily period corresponding to the period of the earth's rotation, and it would have to be so penetrating that its influence would be felt not only inside ordinary laboratory buildings, but also as deep as a salt mine six hundred feet below the surface—as Wahl's experiments on cosmic rays had shown. To von Frisch, that suggested some unknown form of radiation.

What experiment could he devise that would distinguish once and for all between the two?

He began by making a straightforward logical analysis of the facts that lay before him. If he were dealing with some unknown form of radiation, then that radiation would have to have a daily periodicity. More, the periodicity could only be derived somehow or other from the daily rotation of the earth. And in that case, the periodicity of the radiation would have to correspond to the apparent movement of the sun. Therefore, von Frisch reasoned, the intensity of the radiation would necessarily vary from time zone to time zone in the same way as the sun's height in the sky varies from time zone to time zone.

Having made this logical analysis, von Frisch was fully receptive to the sudden flash of insight that followed. All at once he knew the experiment that would answer the critical question.

First he would train a group of bees to seek food at a definite hour by German local time. In a bee room under controlled constant conditions he would then load the bees on a fast transatlantic steamer sailing west. If the bees kept to local German time, they would have shown that their clocks were internal; if they gradually shifted their feeding time as

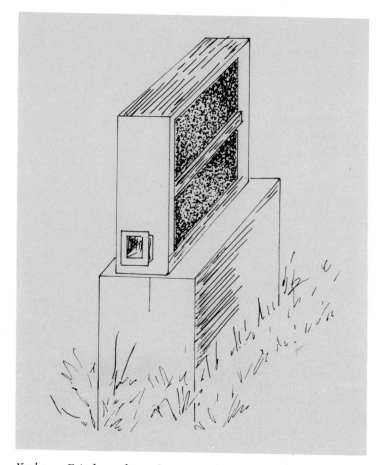

Karl von Frisch used an observation hive like this for his famous displacement experiments. In the drawing the cover on the right is not shown, in order to provide a view of the interior. The flight funnel is shown at the lower forward edge of the hive.

the ship sailed west through time zones, then they would be adapting their timing to some external influence that shifted from time zone to time zone.

Von Frisch made his preparations for the critical experiment with all speed. He began training a group of bees to respond to local German time; he prepared his bee rooms in full readiness to load aboard ship; and he enlisted the aid of an enthusiastic student to make the crossing with the bees and to make detailed notes of their behavior.

Even as patient a man as von Frisch could hardly contain himself as he awaited the day when he could study the many pages of data that his student would bring back.

But she made not a single note.

She was so seasick for the entire voyage that she could not raise her head.

Then came the war. And that project was laid aside for a long time. But later, von Frisch thought that it had probably been a good thing. When he did return to the experiment years later, planes were flying regularly from Paris to New York, and the original plan of the experiment could be much improved.[4]

Meanwhile he reluctantly returned to the experiment that he had been putting off. His reasons for holding back are most human:

"Since the sun's position in the sky changes in the course of the day," he wrote, "either the sun can be used as a compass only for short intervals or, in determining direction, its diurnal course and the time of day must be taken into consideration. Bees know the time of day, but it struck me as fantastic to presuppose that the shifting position of the sun was known and allowed for. Consequently I kept deferring a 'displacement experiment' that was designed to throw light on the matter. When at last we carried it out, there opened before us a view of unexpected new territory."[5]

By the fall of 1949, von Frisch could put off the displacement experiment no longer. His plan was really quite simple. Starting in one location, he would train a group of marked bees (marked with a code of dots of colored shellac, as Beling had done years earlier) to visit an anise-scented feeding station due west of the hive. Then he would move the closed hive to

[4] Karl von Frisch: *A Biologist Remembers* (Oxford: Pergamon Press Ltd.; 1967), p. 178.
[5] Karl von Frisch, *The Dance Language and Orientation of Bees* (Cambridge, Mass.: Harvard University Press; 1967), p. 333.

a new location, and place four identical feeding stations due west, north, east, and south of the hive. Observers would be placed at each station to count the bees that appeared, and, unhappily, to kill each one after it came; von Frisch could not risk the possibility that one bee might learn the proper flight direction from another.

For his first displacement experiment von Frisch chose the village of Brunnwinkl as the training site. There over a number of days he trained marked bees to find a feeding station 220 yards to the west. The flight path took the bees between a few houses, through trees, and over a bay, to reach their goal.

Then on a September morning in 1949, before flight time began, he moved the hive to a broad level field some three miles away, where the bees had never been. The four feeding stands were set up at the four compass points, each 220 yards away from the hive. When the observers were set at each station, von Frisch gave the signal to open the hive, at 8:25 A.M.

He himself was the observer at the west station. Clad in Bavarian field costume, his shock of white hair shining in the morning sun, he settled his tall, rangy frame on a camp stool. Within 20 minutes after the hive was opened, the first bee appeared there. By his own admission he was astonished to see individual bees succeed her, each coming individually. At the end of the first hour, 10 bees had appeared; during the second hour 8 more came; and in the third and final hour there were 2 stragglers. Thus a total of 20 bees had visited the west station—the direction in which they had been trained at Brunnwinkl.

What had the other observers seen?

The one at the north station saw 1 bee; the one at the east station also saw 1; the one at the south station saw 5. Thus out of 27 bees reporting, 20 came to the west station. The 5 at the south station seemed puzzling, but von Frisch suspected that this could have been due to a south breeze that might carry the anise scent to the hive. But aside from that, 20 out of 27 bees reporting to the west station could lead to but one conclusion. The sun guided the bees. Inescapably they had compensated for the sun's movement, for it had been in the west during the last foraging flights at Brunnwinkl, while the next morning in the open field, it had been in the southeast. What von Frisch had thought so fantastic that he put off the experiment had turned out to be the fact.

But this was not all. Later he put another group of bees to a far more severe test. He took an observation hive from Munich ten miles east to the Grub Estate, and there in an entirely unfamiliar location, he permitted the bees one single day of training. He opened the hive shortly before noon, and by midafternoon the bees had found the feeding station that he had placed 200 yards to the northwest. At 8:00 P.M., after the bees had had little more than four hours of training, he closed the hive.

The next morning, before flight time, he moved the hive 14 miles to the Dachau Moor, where the landscape was entirely different. Instead of the woods that lay to the east on the Grub Estate, the waters of the Bagger-See now lay to the east. And instead of dense ground cover and trees, the Moor was bare and uninviting.

Again von Frisch placed four feeding stations at the new location, one to the northwest direction (the direction of the previous day's training), and the others at the three equidistant compass points. To add to the bees' confusion, he turned the hive entrance at right angles from its position of the previous day.

Could the bees now tell the proper flight direction after only one day's training, in a different location, and at a different time of day?

Actually, not all of them did. Of the 30 marked bees that participated in the first day's training, 11 failed to find any feeding station during the three-hour period from 7:15 to 11:15 in the morning.

But of the 19 that did find stations within those hours, 15 went in the correct northwest direction, 2 went 90° to the right, and 2 went 90° to the left; not a single one went completely wrong.

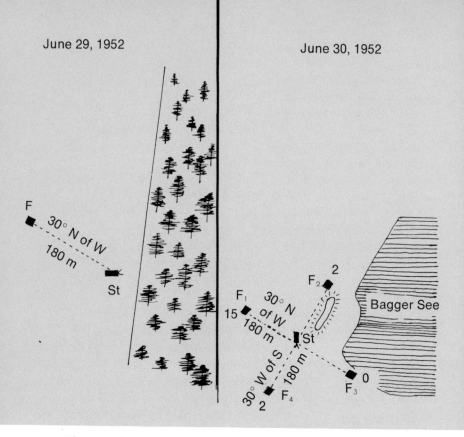

June 29, 1952

June 30, 1952

F

30° N of W

180 m

St

F₁

15

30° N of W

180 m

30° W of S

2

F₄

St

F₂

2

180 m

F₃

0

Bagger See

The two locations—fourteen miles apart—of one of Karl von Frisch's critical displacement experiments. At the Grub Estate, left, the feeding station (F) was placed 180 meters northwest of the observation hive, (St). On the Dachau Moor, right, four feeding stations (F₁), (F₂), (F₃), and (F₄) were placed 180 meters from the hive, (S). As indicated, 15 bees appeared at (F₁), 2 at (F₂), 0 at (F₃), and 2 at (F₄).

"Thus," von Frisch had again to conclude, "it was clear that bees can at any chosen time find a given compass direction according to the sun's position, by taking account of the time of day and the sun's diurnal course. The sun is a perfect compass for them."[6]

[6] *Ibid.*, p. 338.

But again rose the perplexing question that the seasick student had not been able to answer. Was the clock of the bees purely internal, or was it timed by some outside force?

Von Frisch had, of course, considered a number of outside forces, even to such things as atmospheric electricity and cosmic rays, but one by one each of them had been eliminated by careful experiment.

If the search for an outside force as a time-giver was then to be in vain, what internal rhythms were there?

One, the rhythm of hunger, comes immediately to mind. Bees learn very quickly, and having learned, for example, that buckwheat nectar flows only in the morning, they might very easily fall into a morning hunger rhythm. What more natural than that hunger pangs would drive them out to the buckwheat fields on schedule?

However, because of their intimate knowledge of how bees feed, von Frisch and his collaborators thought this an unlikely explanation. In the first place, foraging bees do not use nectar as food directly. Instead they collect it in what is known as a honey bag—an extension of the gullet—and when they return to the hive they turn the contents over to hive bees, who then process it into honey. It is the honey that bees feed on, not the nectar. Moreover, collector bees consume honey when they choose, and this may not be when they are collecting. Finally, bees also consume pollen to add protein to their diet, and the availability of pollen has no relation to the time of nectar flow.[7]

But in spite of its apparent lack of support, von Frisch and his students were unwilling to reject the hunger rhythm idea without putting it to experimental test. When they did, the results were just as they had expected.

"The hunger-cue possibility was investigated with scientific exactitude," wrote Max Renner, one of von Frisch's ablest

---

[7] For a fine modern account of the social life of bees, see Peggy Pickering Larson and Mervin W. Larson: *Lives of Social Insects* (Cleveland and New York: World Publishing Company; 1968), pp. 67–117.

collaborators: "It was found that the time sense of the bees is definitely not governed by a hunger rhythm."[8]

Since, then, research on particular factors—both internal and external—had so far yielded negative results, the time had finally come to attack the more general question. Whatever its actual source, the clock was either internal or external. Which was it?

On his second trip to the United States in 1949, von Frisch had flown from London to New York, and had been impressed by the speed and convenience of the trip. By 1955 fast flights across the Atlantic were commonplace, and von Frisch decided that the time was ripe to return to the experiment that sea-sickness had defeated long before.

As before, the object of the experiment was to find out whether bees trained to feed at a particular time in Europe would maintain that time on arrival in the United States, or if some external force would shift their timing.

Max Renner shouldered the chief responsibility for the program, and von Frisch made the necessary preliminary arrangements through his worldwide scientific contacts. Plane connections were most favorable between Paris and New York, and Professor Grassé of Paris offered the use of his laboratory for training the bees to French time. Professor T. Schneirla arranged for the use of a room in the American Museum of Natural History in New York.

The essential equipment was contained in two identical bee rooms, in which the bees would be kept at all times in constant light, temperature, and humidity; and shielded from any clue whatever as to external events. The rooms were built in Munich, made collapsible for easy shipment to Paris and New York. Upon assembly at their destinations, they were some 8 feet wide, 9 feet high, and 15 feet long. Besides the table on which the beehive would rest, each room was provided with identical pollen pans, drinking troughs, feeding tables, and stools for the observer, who naturally sat inside near the

---

[8] Max Renner: "The Clock of the Bees," *Natural History*, Vol. LXVIII (October 1959), p. 438.

Max Renner used a bee room like this to train bees in Paris. He then flew them overnight to an identical bee room in New York. The equipment included lights, (a); transparent paper, (b); geometric figures, (c), for the bees to orient themselves by; beehive, (d); thermometer and relay, (e); electric heater, (f); pollen pan, (g); humidity regulator, (h); drinking trough, (i); and feeding table, (k). Renner sat near the feeding table.

feeding dish. Identical geometrical figures were painted on the walls to aid the bees in orienting themselves.

Finally on a June day in Paris in 1955, Max Renner began training a group of 40 bees to report to the feeding dish between 8:15 and 10:15 A.M. French Summer Time. By June 13 he concluded that they were adequately trained and, packing them into a box, he flew with them to New York. Professor Schneirla was waiting at the airport to help speed the bees through customs, and in less than 24 hours after they had left Paris, the bees occupied an identical room in the American Museum of Natural History.

How would they answer the question first asked so long ago?

Would they start feeding just 24 hours after their last training in Paris?

Or would they wait out the 5-hour time difference between Paris and New York to receive a clue from some external force geared to Eastern Daylight Time?

Renner said afterward that he sat "expectantly" near the feeding dish. "Expectantly" may be the understatement of his career.

"The bees did not take long to answer these questions," he wrote, "and they answered in the clearest possible way. At 3:15 EDT [that is, just 24 hours after Paris training time], the first bees came out of the hive and started flying about the room, as if their location had never been changed. And the ensuing visits to the feeding place were so numerous and thorough that it was difficult for the observer to note down each visitor correctly.

"The result of this experiment, as well as the reverse one (training in New York and testing in Paris), has clearly answered the question as to the nature of the bees' orientation in time: the trained collector bees maintained their 24-hour rhythms independently of external influences that periodically recur during the day. Bees have an *internal* clock, governed by their organism."[9]

[9] *Ibid.*, p. 439.

So human a man as Karl von Frisch could not fail to have been delighted by the eminently successful outcome of Renner's work. Did he then regard the clock of the bees as fully understood?

To close his autobiography, in which he gave the highlights of his half century of work with bees, he said:

"Not that we had solved thereby the last mystery of their time sense. By no means. Science advances but slowly, with halting steps. But does not therein lie her eternal fascination? And would we not soon tire of her if she were to reveal her ultimate truths too easily?"[1]

[1] Karl von Frisch: *A Biologist Remembers*, p. 178.

A small beaste they haue,
they call Assapanik,
but we call them flying squirrels,
because spreading their legs
and so stretching
the largeness of their skins . . .
they haue bin seene to fly
30 to 40 yards.
CAPTAIN JOHN SMITH,
1624

11. DeCoursey: Treadmills and Flying Squirrels

For a fresh view of modern research on living clocks, let us now look to the younger biologists whose work is already very highly regarded. High among them we find Patricia Jackson DeCoursey.

When Professor Peter Marler, noted authority on animal communication and behavior—now at Rockefeller Institute—was at the University of California, I asked him to name a few promising young researchers working in the field, and he answered without hesitation.

"Right off, you'd have to start with Pat DeCoursey. She's young, very pretty by the way, and very, very sharp. How many people d'you think have ever been asked to present Ph.D. thesis work before an international audience?" He spread his hands. "She was, in 1960 . . . still under thirty. Now every rhythm book you come across cites her symposium paper as a major reference."

How might such an unusual development come about? For perspective let us go back to the late 1930's, and make our way into the Quetico-Superior wilderness area that lies astride northern Minnesota and Ontario. Into that area let us place a little girl of three, fascinated by wild things, and gifted with a rare capacity for patient observation.

Two hundred miles above Duluth the Pigeon River empties into Lake Superior. Near Grand Portage below the river's mouth, the French-Canadian voyageurs of the eighteenth-century laid in provisions for the journey from the North West Company Post into the vast wilderness. Up wild rivers, over rough portages, and on across countless lakes, they sought the Northwest Passage and came back laden with furs of beaver, fox, marten, and mink.

Two decades after the first voyageurs had passed that way, the Province of Ontario dedicated Quetico Park—eighteen hundred square miles of forest wilderness lakeland—to ensure the preservation of the old canoe country in its original state. Today one may point a canoe up the Pigeon River, follow the Grand Portage Route toward Saganago, and very soon find himself in restful tune with his environment.[1]

The country is one of wild and ever-changing beauty. Hills dark with ancient pine slope upward from the rivers. Fragrant balsam and cedar crowd the valley floors, which are dotted here and there with maples and with birches large enough to be stripped for canoes. Along the lakeshores, fissured cliffs thick with mosses, lichens, and ferns fringe the water. Where the rivers widen to flow lazily over the marshlands, ducks and beaver and moose move undisturbed by man. It is a serene world, and the clear air carries the calls of birds,

---

[1] A modern "voyageur's" journey along the Grand Portage Route has been entertainingly described by J. Arnold Bolz: *Portage Into the Past: By Canoe Along the Minnesota-Ontario Boundary Waters* (Minneapolis: University of Minnesota Press; 1960).

the rustle of animals moving through cover, and the distant plash of swift rivers foaming through steep gorges.

For men, it is a land of soul-washing peace. For wild creatures, it is a free land of water, earth, air, and sky, in which each may find the place best suited for his own survival and that of his species.

Into the heart of the Quetico wilderness, year after year, came a young Wisconsin doctor and his wife to share with their small daughter Patricia the delights of watching beavers at work, and bear and deer and bristly porcupines, and moose surging shoulder-deep through quiet waters.

Here the doctor's family could scent the marsh grass, and listen to mallards quacking in the rice beds. They might watch as the ducks rose on purple-barred wings, droplets of water glistening from iridescent green heads and dripping from webbed feet. Near a pond they might be fortunate to hear a dissonance of squeaks, squeals, grunts, and moans, and peering through brush and cattails, watch a colony of beaver at play. Swimming and diving, the beaver would slap their tails on the water, swim round and round, then tiring, turn over on their backs and float lazily.

Through feathery groves of cedar the family could stroll, the balsam-scented air vibrant with the sounds of birds: nuthatches yanking, chickadees chinking, and grouse whirring aloft. Redcrested woodpeckers eighteen inches long might be seen chipping nest holes in old snags. Overhead, red-tailed hawks soared, and fluffy Canada jays glided silently against a background of scarlet rose haws and blue asters.

It was here that Patricia Jackson learned, from the age of three on, the shapes and colors and flight patterns of the Quetico birds, and the sounds of their calls and songs. Years later she would read in books that "nowhere in the animal kingdom has sound reproduction become as highly perfected or as widely used as it is among birds," and further that these sounds, calls, and songs serve reproductive functions, social functions, and individual functions.[2]

[2] Joel Carl Welty: *The Life of Birds* (New York: Alfred A. Knopf, Inc.; 1963), p. 188.

By the time she was ready for high school in Madison, Wisconsin, Patricia had given up her earlier intention to study medicine, for she had known for some time that the study of animals was of far greater interest to her. While still in high school she did a few "little research projects" with birds.

Next came the choice of a university where she might find the best undergraduate preparation for a research career in biology. She chose Cornell because of its fine tradition of liberal education and its faculty of able men in physics, chemistry, and the life sciences. At Ithaca, Patricia Jackson gave her major attention to physiology, with special emphasis on the functions of the endocrine system.

As she knew, most of the processes that occur in living organisms are regulated by two mechanisms: those that take pace through the nervous system, and those that are governed by chemical regulators. She was most interested in the chemical regulators, those substances—chiefly hormones—that are produced by the endocrine glands: the pituitary, the thyroid, the parathyroid, the islets of Langerhans, the adrenals, the gonads, and the placenta.

And she continued this interest when she returned to Madison to register as a graduate student at the University of Wisconsin. She earned her Master's degree in 1955 with a thesis on biochemical endocrinology, and she began to survey fields from which to choose a subject for her Ph.D. research.

Her freedom of choice was practically unlimited. Wisconsin is noted for its devotion to the principle that if graduate students are to mature into independent workers, they had better start their independence early. In this atmosphere Patricia Jackson DeCoursey (she had married biologist George DeCoursey in 1954) thought back over all of the animals that she had known since childhood, and decided that she wanted to work with the flying squirrel, *Glaucomys volans*.[3]

This was a logical choice for several reasons. For one thing, flying squirrels were numerous near Madison, and their habit of returning to the same den tree every day made them

---

[3] The genus-species name means "silver-gray mouse that flies."

easy to keep track of. For another, they could be studied in their natural state, or the next thing to it, and Patricia De-Coursey wanted to study living animals—not their endocrine organs macerated and extracted in a flask. So she began studying the daily activity patterns of flying squirrels in the woods, and very soon was fascinated by the precision with which they timed their activities.

Then a coincidence occurred that was to have a lasting influence on her future work. At Cornell she had known Kenneth S. Rawson, who at that time was an instructor, and who was soon to write his own Ph.D. thesis at Harvard on "Homing Behavior and Endogenous Activity Rhythms."[4] Now Rawson arrived at Madison and lectured on the rhythms of mice. "This single lecture," Patricia DeCoursey says, "had an immense impact on all of my subsequent ideas and experiments."

Mice and flying squirrels belong to the same order—Rodentia—and Rawson's results with mice triggered a host of questions in DeCoursey's mind, beginning with "What would happen if . . . ?"

Of course the observation that flying squirrels time their departure from the den tree indicated only that they ought to be first-rate animals to use for experiments under controlled conditions.

DeCoursey's first concern after that was to find laboratory space where such conditions could be maintained, and she prowled the Madison campus for rooms in which the temperature could be held constant and the level of light could be controlled. Also she needed a second area for breeding and maintaining her colonies.

For the colonies, she found an entire building. But what a building! Ancient, derelict, and at the outer edge of the campus, it was still, for some vague reason, called the Behavior Lab. But it turned out to serve well; it already had breeding cages, and with energetic cleanup and rearrangement it met the

---

[4] Most Ph.D. theses lie buried in archives. Rawson's, however, completed in 1956, has been widely cited in the professional literature.

essential requirements. For her experimental work, DeCoursey was assigned several sub-sub-basement rooms in the Biochemistry building. Uninviting as they at first seemed, they had two essential features: even temperatures in spite of Madison's sultry summers and frigid winters, and freedom from disturbance.

In one of these underground rooms DeCoursey built all of her equipment with her own hands from scratch. She installed supplementary temperature controls, and set up a lighting system that would operate automatically and continuously to provide a range of light-dark cycles. She built walk-in cages, and delicately-balanced wheels for her squirrels to run in. She connected each running wheel to a continuous electrical recorder to provide 24-hour charts of the activities of her animals. Later she would mount the daily charts in vertical series to give concise activity graphs over long periods.

After long days of wrestling heavy lumber, sawing, fitting, and nailing, she would come home utterly exhausted, and there were times when she told herself that she was working for a degree in carpentry.

But finally all was ready. Now she could get down to real work. During the five-year period from 1955 to 1960 she often toiled through experimental periods that demanded her attention throughout seven-day weeks from dawn into the small hours of the morning. Breakdowns of equipment happened with exasperating frequency, meaning even longer hours. Crises required spur-of-the-moment action, and some of her stop-gap remedies worked so well that she used them routinely from then on. During each day's work, she planned the experiments to follow, and made modifications of her equipment so that she could go on without losing time.

And every day she had to trudge across campus to the Behavior Lab to care for her colonies. Workers with animal rhythms learn early that their subjects have no sensitivity to human fatigue.

Let us now look at some of the remarkable results that Patricia DeCoursey achieved—results that the international

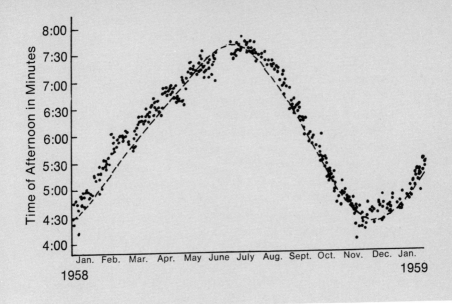

This outdoor running-wheel experiment shows how precisely flying squirrels begin their evening activity. The dashed line is the time of local sunset. The dots represent the times of onset of running activity. Note that the record covers more than a full year; it was produced by a single carefully tended animal.

audience of experts at Cold Spring Harbor found so impressive, especially from a young woman just finishing her Ph.D.

First let us see how astonishingly precise she found the rhythms of flying squirrels to be. In her first field observations she found that flying squirrels leave their den trees close to dusk.[5]

Now she began recording the rhythm of activity of a single squirrel in a recording cage set up outdoors. When the squirrel first stirred, he almost immediately began running on his wheel, and the electrical recorder drew closely spaced lines

[5] At exactly what time does "dusk" arrive? One way to pinpoint it would be to define dusk as that time of day when the natural light intensity falls to some arbitrarily specified level in foot-candles. One of the first questions that DeCoursey wanted to answer was "exactly what light intensity is associated with a flying squirrel's onset of activity?" Ten years later too many problems of far greater interest had come up to give her time to find the answer to that one.

on chart paper corresponding to the activity moment by moment. From these charts DeCoursey read the exact time of afternoon at which the squirrel began to run.

She started her recordings in late December of 1957 and continued them through January of 1959. As the charts unwound she plotted the results in comparison with the exact time of local sunset, and found a remarkable correspondence. This close relation between onset of running and dusk naturally suggested that some condition in the environment must regulate the onset. Or, to put it the other way around, the animal received external clues that enabled him to time himself with high precision.[6]

What would happen then, if the clues were entirely removed? To find out, DeCoursey took some of her squirrels to her sub-sub-basement where she could shield them from all outside information. In constant conditions of 20° centigrade (or 68° Fahrenheit) and complete darkness, each squirrel established a rhythm peculiar to himself—the so-called "free-running" rhythm. Even though removed from all light and temperature clues, the squirrels continued to alternate strong running on the wheel with periods of rest; each period, moreover, took approximately—but not exactly—24 hours, thus drifting out of phase with the day-night changes in the outside world.

DeCoursey used the time of the abrupt onset of activity as the reference point for each circadian period, and from this she measured the rate and direction of the drift. In all, she used 18 different flying squirrels for 48 different test periods lasting all the way from 10 to 123 days. The periods shown by different individual squirrels ranged from 22 hours 58 minutes to 24 hours 21 minutes. And notwithstanding the differences between individuals, the period of each particular animal after it had adjusted to "running free" was constant within a very few minutes.

---

[6] Alert readers will notice a short gap in June of 1958 between the dots in the accompanying figure; they will agree that if this represents a short vacation, it was well earned.

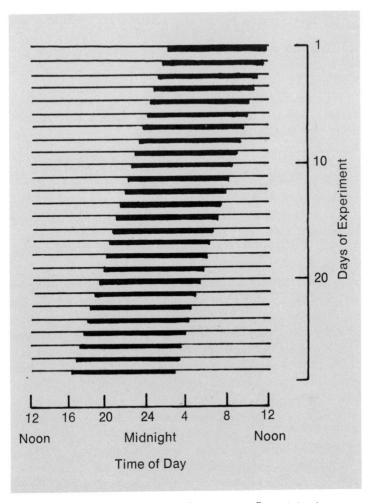

A twenty-five-day record of one flying squirrel's activity in a running wheel in continuous darkness. Deprived of the normal light-dark clues as to the time of day, this animal gained continuously on the clock.

"Thus the outstanding characteristics of the activity rhythms in darkness," she wrote, "appeared to be (a) their persistent and apparently endogenous nature, (b) the stable frequency for one individual during one test period, (c) the restriction of the free-running rhythm to a narrow range close to and usually less than 24 hours, with a resultant drift of the activity period, and (d) the very high accuracy of the rhythm of an individual."[7]

Well and good. But only the beginning. In our own environment—considering men purely as animals for the moment—the factors that affect us most strongly are light, temperature, and sound, and it would seem logical that the same would be true of other animals. In assessing the relative importance of these factors on setting the rhythms of flying squirrels, it seemed to DeCoursey that light should have the most influence, because light intensity is clearly associated with the activity cycles of almost all animals, no matter whether they are day-active, nocturnal, or crepuscular, that is, active only during the dim light near dawn and dusk. But she kept the possible effects of temperature and sound in the back of her mind, and planned to make some tests using them also.

First, how might she design experiments to see more clearly just how light may control activity? One plan would be to place animals in constant laboratory conditions like those she had used to measure free-running rhythms; and following that, to subject the animals to alternating periods of light and darkness. If a change occurred following such an alternating light-dark cycle, the experiment could then be closed off by returning the animals to the original constant conditions to see if the original free-running rhythms would be resumed.

DeCoursey explored this in 30 different experiments with squirrels whose free-running rhythms she had already found to be both longer and shorter than 24 hours. In a typical experiment, she placed a squirrel whose free-running rhythm was less than 24 hours in constant darkness for several days. As she expected, the squirrel's activity onset drifted forward with respect to the clock (that is, it ran faster than the clock). Next she exposed the squirrel to alternating periods of 12 hours of

---

[7] Patricia DeCoursey: "Phase Control of Activity in a Rodent," *Cold Spring Harbor Symposia on Quantitative Biology*, Vol. XXV (1960), p. 49.

light and 12 hours of darkness, which were out of phase with the free-running rhythm.[8] Now the onset of activity drifted backward (that is, lost time), until it reached a constant value that persisted as long as the light-dark schedule was maintained. To end the experiment she returned the squirrel to constant darkness again, and its activity onset immediately resumed its forward drift.

In further experiments she used animals with free-running rhythms longer than 24 hours, and also different light-dark cycles, such as 10 hours of light alternating with 14 hours of darkness. The results she obtained were beautifully consistent, and added further support to the view that light is a phase-controlling agent.

Still she was far from satisfied, because these experiments did not really pinpoint the effect of light. So she took up a second method in which she used short isolated single exposures of light—a kind of light-shock technique, as it were.

In one of her sub-sub-basement rooms she again kept a flying squirrel in constant darkness for a number of weeks, again to check its free-running rhythm in the steady state. This animal had a free-running rhythm of less than 24 hours, and again the onset of running drifted forward with respect to the clock, as it had before. Now, however, she continued the experiment for enough days to permit her to draw a sloping straight line across the daily onset times; having established the slope of the line, she could then predict by extrapolation the time that onset would occur on the next day.

Having thus established the rate of forward drift of the free-running rhythm, she exposed the squirrel to a light-shock of *half a foot-candle* for 10 minutes *once only* at one point in a

---

[8] She used a light intensity of one foot-candle. The sensitivity of flying squirrels to light can be judged from the recommendations that lighting engineers make for various industrial situations: 100 foot-candles for small detail and high-speed operations; 20 foot-candles for ordinary working conditions; and 5 foot-candles for casual passageways that are free of obstructions.

24-hour period, and measured the difference between the time of the next actual onset and the time she predicted from the slope of the drift line.

She found that the single light-shocks affected onset at certain times, and at other times they had no effect at all. Whether or not a phase shift followed at light-shock *depended on what time of day the squirrel thought it was!*

Consider a flying squirrel whose own free-running clock runs on a 23-hour day. He will then gain an hour a day on a standard clock. For him, dusk will come every 23 hours; the night will last about 10 hours, and day about 13. We may think of the 10-hour period as the squirrel's subjective night, and the 13-hour period as his subjective day. Consequently DeCoursey gave the light-shocks at a number of different times during the squirrel's subjective night as well as during his subjective day to see if any differences would appear.

She found the most marked effect when she gave the light-shock just as the squirrel's free-running clock was telling him that dusk was at hand. At just that time, the light signal over-rode the free-running time signal. It was as though the squirrel had thought: "Wup! Not dusk yet. Better lie low."

However, when the light-shock was given some time after the squirrel had become aware that it should be night, and was fully active, it affected him far less. Moreover, if the shock was given during the day when he was inactive, he paid no attention to it at all.

Thus DeCoursey's experiments with single isolated light-shocks helped her to clarify her understanding of the mechanism by which flying squirrels synchronize their movements when in the wild state with the time of day, from one end of the year to the other. This synchronization could be regarded as a series of daily *resetting* phenomena, which were associated with a light-response rhythm, with light being the phase-controlling agent.

Following this, DeCoursey carried out experiments to see what phase-controlling effects temperature and sound might have. However, neither temperature-shocks nor sound-shocks had any effect at all.[9]

These, then, are only a few highlights selected from five years in the research life of a young biologist. Patricia DeCoursey put it in one paragraph.

"In summary, it may be said that in *Glaucomys* the activity in natural conditions coincides with the period of darkness, and this relationship is maintained throughout the course of the year by the interaction of an endogenous activity rhythm and a related daily rhythm of responsiveness to light. Such a resetting rhythm could not be demonstrated to a variety of temperature and sound stimuli."[1]

Since that writing, Patricia DeCoursey has been an honorary research fellow at the University of Wisconsin, a research collaborator at Washington State University, and an instructor and research worker at the University of South Carolina. Looking ahead to the decade of the 1970's, she would not change anything that she did to prepare herself for it.

"The scientific years have been golden years," she says, "stimulating beyond measure, and my only hope is to be able to challenge intellect and fingers to even-further solutions to biological riddles."[2]

Whatever future direction she follows, her work will

[9] Charles Darwin once tried the effect of sound on a sensitive plant. In recalling the incident his son Francis wrote, ". . . he was willing to test what would seem to some people not to be worth testing. These rather wild trials he called 'fool's experiments,' and enjoyed extremely. As an example I may mention that finding the seed-leaves of a kind of sensitive plant to be highly sensitive to vibrations of the table, he fancied that they might perceive the vibrations of sound, and therefore made me play my bassoon close to a plant. This is not so much an example of superabundant theorizing from a small cause as of his wish to test the most improbable ideas." (Charles Darwin: *The Autobiography of Charles Darwin, and Selected Letters*, ed. Francis Darwin [New York: Dover Publications, Inc.; 1958], p. 101.)

[1] Patricia DeCoursey: "Phase Control," p. 53.

[2] Patricia DeCoursey to Ritchie R. Ward, personal communication, September 20, 1969.

continue in the field of the physiology of animal behavior—
"the field that relates the physiology of the whole, intact animal
to the environment in which he lives"—an environment reminiscent, perhaps, of the Quetico-Superior wilderness area that
she loved as a child.

12. Sauer and Sauer: Celestial Navigation by Birds

If scientists were amazed by Gustav Kramer's discovery that birds orient themselves by the sun, they were truly astounded a few years later to hear evidence that night-flying birds navigate by the stars. That idea was advanced by E. G. Franz Sauer, of the University of Freiburg, Germany, and his wife Eleanore.

As has happened with so many biologists who have found themselves caught up in the study of living clocks, Franz Sauer and his wife were also drawn into the field from another direction—through work that was aimed at understanding more about how animals learn.

Franz Sauer had taken his doctorate at Freiburg in 1953, in zoology, but he found that he was especially interested in the problems of animal behavior, particularly as they are affected by the environment in which the animals live, and he went on to take a second doctorate at Freiburg in the specialty of ethology—if specialty it may properly be called, for it ranges over all of the geographical regions and climates of the world.

Thus, while we might find a physicist or a chemist spending his early professional years in a single laboratory, we find Franz Sauer studying warblers in Europe, animals of the temperate zone in Wisconsin, golden plovers in California, subtropical life forms in Florida, and ostriches in South Africa.

But whatever animal species he studied, a particular phase of animal behavior that Sauer found most fascinating was the phenomenon of learning, and the communication that is associated with it; and an early problem to which he and his wife devoted the utmost care and unbelievable patience was how young birds learn to sing.

This was a very controversial point in the early 1950's. Many zoologists thought that birds learn to sing only as a result of hearing and then imitating the songs of older and more experienced companions. And it was certainly in accord with other observations of how animals learn. Anyone who has watched a mother quail, with limitless patience, teach her brood of tiny fluffballs how to hop down a set of garden steps, would find it hard to believe that bird behavior is not taught by the adults and learned by the young: would find it especially hard to believe that bird song—having, as it does, so many different variations even in a single species—is not learned.

But that conclusion relies on "what stands to reason," and it often turns out that what stands to reason is not so.

The Sauers set out to find out what *was* so, for bird song.

The classical method of research, of course, is to start with a hypothesis—a working assumption—and then to test that assumption experimentally to see if the observed facts seem to bear it out. That is the method that the Sauers chose, knowing full well the endless hours of drudgery that lay ahead.

Their working hypothesis was that birds do not really need to learn to sing—that they sing as a result of an inborn natural ability, no matter whether accomplished songsters are there to teach them.

To test this hypothesis, the Sauers began raising young European warblers from eggs, in an aviary in Freiburg. This would have been an onerous task even under ideal conditions, and theirs were certainly far from ideal. An adult warbler weighs less than an ounce, and the newly hatched chicks are tiny indeed, delicate, nervous, and with voracious appetites for live insects. Nevertheless, with extraordinary patience, the Sauers became highly skilled at mothering and raising these delicate little bundles to full adulthood.

More, to meet their experimental requirements, they raised their warblers in a completely soundproof aviary, so that throughout their lives they heard no song from any other bird.

How would these isolated warblers react when they were fully grown, and the mating season was at hand? Would the adult male sing the particular song by which he would announce to all that he had staked out a territory and was ready to raise a family?

Would he, in fact, sing at all?

As we saw earlier, the great Karl von Frisch used to tell his graduate students at Munich that they would have to look a long time before they ever found a biological experiment whose results were completely quantitative. So too it was with the Sauers and their warblers. Their results were not airtight. Nevertheless, on one critical day they were overjoyed to find that all their labors had been rewarded.

A warbler sang!

Although he had never heard another bird sing in all his life, he sang. Then another isolated warbler sang, and another. It was hardly expert warbler song—that would have been too much to hope for. But it was clearly identifiable as the proper song for that warbler to sing at that time in his life.

Of course the Sauers might have hoped for stronger statistical support than their final results showed, but they had successfully tested the hypothesis that they had set out to test, and that was a triumph.[1]

[1] Donald R. Griffin: *Bird Migration* (Garden City, N.Y.: Doubleday and Co., Inc.: Natural History Press; 1964), p. 139.

Even so, what an advantage the physicist or chemist has over the biologist! If a nuclear physicist discovers a new radio-active element and describes his experimental procedures carefully, then any other physicist anywhere else in the world—given comparable experience and equipment—can duplicate the result. Not so, always, with the biologist. He lives by approximations—by what is probably so—even though today he can make a very good statement of the statistical significance of his observations.

But as compensation, a skilled biologist often acquires delicate experimental techniques that may be virtually unique, and he can sometimes use them in daring ventures that no other scientist would have the temerity to attempt.

This was precisely the situation in which the Sauers found themselves in the middle of the 1950's. They had been en-thralled by Gustav Kramer's discovery that starlings orient themselves by using the sun as a compass.

But that only presented a new question.

Starlings migrate during the daytime, when the sun presents a ready reference. What then of the birds that migrate only at night? Do they depend on the stars? Incredible!

Yet it had been known for years that many species of birds migrate mostly at night, and that over very long distances. The warblers, with which the Sauers had developed such great experimental skill, belonged among these species. What more exciting, then, than to undertake to shed new light on problems that had been so brilliantly attacked by "our good old friend, the late Gustav Kramer, most honored pioneer in the study of bird navigation."[2]

So began the Sauers' intensive study of how warblers navigate by the stars. In the wild, these small songbirds in the spring throughout Europe, raise their young, and during mid-summer feed contentedly on elderberries and blackberries. But on an appointed night in August, the entire resident population

---

[2] E. G. Franz Sauer to Ritchie R. Ward, personal communication, March 3, 1970.

of one species will disappear. The southward fall migration will have begun. The next morning a new group—perhaps of the same species—will have arrived from a more northern area, to stay a few days, and then follow the southward flyway.

Throughout September and October, as the Sauers observed, there was a continual arrival of warblers from the north, followed a few days later by their departure for the south. The first to disappear completely were the garden warblers, then the whitethroats, followed by the lesser whitethroats, and finally the blackcaps.

These European warblers, like most small birds, feed by day and migrate by night, when, according to Joel Carl Welty, "they are safer from predators and . . . the air is generally more stable. Small birds, with their livelier metabolism, exhaust their energy stores in flight more rapidly than large birds do. Therefore they must replace them quickly and effectively, and this can be done best by daylight. Furthermore, by migrating at night they are less likely to be distracted by their surroundings and are better able to concentrate on covering distance."[3]

Where do the warblers go for the winter? This has been well established by banding the birds in Europe and recovering them in their winter homes. In general, they fly to various parts of Africa. The lesser whitethroats, for example, winter in that part of central Africa that ranges from about 10° east longitude to 50° east. Other warblers travel from as far as Scandinavia to southern Africa—a third of the way around the globe. Of course in the spring, the migratory restlessness overtakes them again; some innate sense of the season directs them to prepare for the flight back to the breeding grounds in the north.

"Most remarkable of all," wrote Sauer, "is that each bird finds its own way to its destination! The warblers do not follow the leader or make the journey as a group; they navigate

[3] Joel Carl Welty: *The Life of Birds* (New York: Alfred A. Knopf; 1963), p. 465.

individually. And young birds making their first migration reach their goal as surely as experienced travelers. Somehow, purely by instinct, the warblers know exactly how to set their course."[4]

We have already seen in Chapter 9 how the arctic tern nests within a hundred miles of the North Pole and in the fall flies over Canada, across the Atlantic, and down around the Cape of Good Hope to winter feeding grounds below Port Elizabeth. And how the New Zealand bronzed cuckoo flies from its summer home "down under" to pinpointed wintering areas in the Bismarck Archipelago and the Solomons. These are marvels of migratory flight.

What then of the astonishing feats of the homing behavior of birds? Werner Rüppel took starlings from their nests near Berlin and carried them in various directions as far as 1,250 miles. After they were released, they found their way home! G. V. T. Matthews flew a Manx shearwater by plane from the west coast of England to Boston, where he released it. Covering 250 miles a day, the shearwater flew home across the Atlantic, and found its own nest within 12 days!

We have seen, too, how many different explanations have been proposed for such incredible performances: the birds' ability to sense variations in thermal radiation or of magnetic field strength, as well as of differences in forces caused by the earth's rotation. But all of these fail, in one way or another, as acceptable explanations.

What would be more natural, then, than Franz and Eleanore Sauer taking as their working position the hypothesis that some kind of instinct for navigation—perhaps even similar to man's conscious use of compass, sextant, and chronometer— might lie close to the truth?

Their good friend Gustav Kramer had already shown that starlings can orient themselves by relating the position of the sun to the time their internal clocks tell. The Sauers themselves had shown that warblers can do the same. If birds can orient

[4] E. G. Franz Sauer: "Celestial Navigation by Birds," *Scientific American*, Vol. CXCIX (August 1958), pp. 42–7.

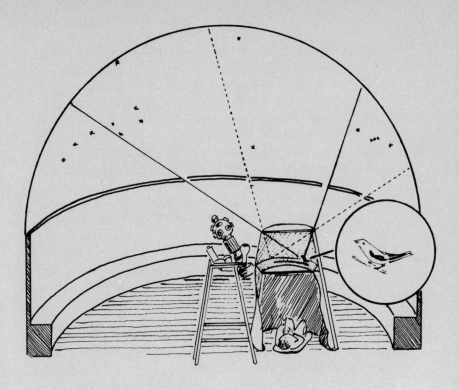

A planetarium experiment is performed under a twenty-foot dome. Dome and bird cage are shown in cross section. A cloth from the bottom of the cage to the floor cuts out light from below. The sector of the sky visible to the bird is indicated by solid lines. The sector visible from the opposite side of the perch is shown by broken lines. "Flight direction" means the direction in which the bird faces.

themselves by the sun by day, why not by the stars by night? This the Sauers set out to test in their aviary at Freiburg.

In their first series of experiments they used a cage with a glass window at the top, so that the warblers could see some of the sky (about 70° of it), but none of the surrounding physical landmarks. In much the same way as Kramer had done with his starlings, the Sauers observed the direction that the warblers favored by observing from below the directions in which they pointed or fluttered during their "flight fidgets."

In an impressive paper dedicated to Dr. H. O. Wagner, of the *Überseemuseums* in Bremen, Franz Sauer reported the details of the observations that he and his wife made.[5]

The directions that the birds favored were in every case characteristic of the species. In the fall, the garden warblers, the whitethroats, and the blackcaps all pointed toward the southwest; this is just the direction they take in their normal fall migration—from Freiburg, say, southwest to Spain, down to Gibraltar, and across the Straits to Africa.

The lesser whitethroats, on the other hand, fly from Germany southeast across the Balkans, and then turn south up the Nile River valley. In their cages in Freiburg, they also pointed southeast—toward the Balkans. Moreover, it did not matter whether a particular bird was an experienced navigator that the Sauers had trapped in a thicket in Freiburg or a warbler that they had hand-raised in their isolation chamber.

In covering a wide range of natural situations, the Sauers also found that the warblers were disturbed by moonlight, as well as by bright shooting stars. When clouds began to filter across the sky, the birds were at first uncertain, but as long as they could see the brightest of the stars through the veil of clouds, they pointed in the right direction. When the clouds became so dense, though, that stars were no longer to be seen, the birds fluttered aimlessly for a while, and then went off to sleep.

Next, the Sauers placed their warblers in a closed room which they could illuminate either with evenly diffused light, or with polarized light—that is, with light that vibrates chiefly in one direction. In both situations the warblers were completely disoriented.

This, then, was the groundwork for the amazing discoveries that were yet to come. Under the natural night sky, of course, the Sauers had had no way of varying the positions of the stars

[5] E. G. Franz Sauer: "Die Sternenorientierung nächtlich ziehander Grasmücken," *Zeitschrift für Tierpsychologie*, Vol. XIV (August 1957), pp. 29-70.

at will—that is, of changing the positions of the stars to correspond to those of other latitudes or time zones. But this could readily be done in a planetarium, and if the warblers sensed the "stars" in the planetarium as corresponding to real stars, the Sauers could, in effect, place their birds anywhere, without moving them from Freiburg.

Soon, then, through the cooperation of Kapitän M. Berger and Dr. W. Stein of the Olbers Planetarium in Bremen, they were authorized to use a Zeiss planetarium that had a twenty-foot dome. Inside it, they placed a cage for a warbler, and as before, they made their observations of the flight direction from below.

In a typical experiment in which the dome of the planetarium was lighted with diffuse light, the warbler pointed randomly in many different directions, as shown in the 360° circle at A in the accompanying figure. When a blackcap was shown a simulated spring sky, it pointed to the northeast, just as it would under natural conditions (circle B). Under a simulated autumn sky it headed southwest toward Spain (circle C). Finally, a lesser whitethroat, that would be expected to point southeast toward the Balkans, did just that (circle D).

But the star performer in the planetarium was one individual lesser whitethroat. This is the species that follows two successive compass directions in the course of its fall migration —first southeast from Germany across the Balkans, and then due south toward the Nile headwaters. It was, therefore, an ideal subject for experiments at the various latitudes and longitudes that could be simulated in the planetarium.

The individual whitethroat that the Sauers now placed under the planetarium sky was one of those that they had raised from the egg; it had never been outside a cage, much less on a flight to Africa. Its behavior in the planetarium would therefore be completely independent of any previous experience in navigation.

The Sauers placed this lesser whitethroat in the planetarium and adjusted the sky so that it corresponded to the latitude of Freiburg, or 48° north. Confirming their previous observa-

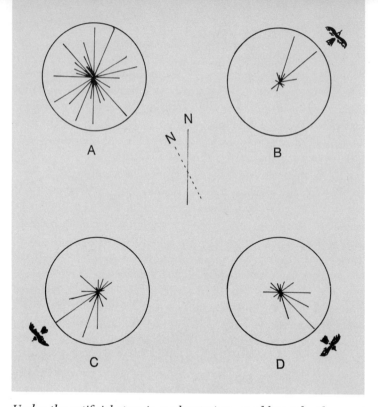

Under the artificial stars in a planetarium, warblers take the same directions they would take in the wild. (A) Under diffuse lighting with no stars visible, the birds faced in all directions. (B) Blackcap under a spring sky. (C) Blackcap under an autumn sky. (D) Lesser whitethroat under an autumn sky. The (N)-(N) directional lines show small differences between true north and planetarium "north."

tions of other birds under the open sky, the Sauers found that this lesser whitethroat faced southeast.

Next they shifted the planetarium sky so that the warbler's apparent position moved farther and farther south. The bird continued to bear southeast until it arrived at a latitude somewhat below 40°. There it began to shift its direction toward south southeast until, at a latitude of 15°, it bore due south. Guided only by images of stars projected on an artificial sky, a bird that had never been outside a cage pointed in the directions required for a flight from Freiburg to the headwaters of the Nile!

It is not, perhaps, too hard to imagine how a bird might be able to sense changes in latitude, either under the natural sky or under an artificial one. A human navigator finds his latitude by measuring the altitude, or angle, that a fixed star such as Polaris makes with the horizon. It would not be unreasonable to suppose that birds determine their latitude in a similar way.

The fall migration pattern of the lesser whitethroat, shown by the fat black arrows, has long been established by banding observations. When a caged whitethroat in Freiburg, Germany, is permitted to see nothing but a planetarium "sky" corresponding to the appropriate latitude, it points in the direction of the natural migration pattern; the small arrows are projections of the directions the birds choose in the planetarium. The hatched area shows roughly the known wintering grounds.

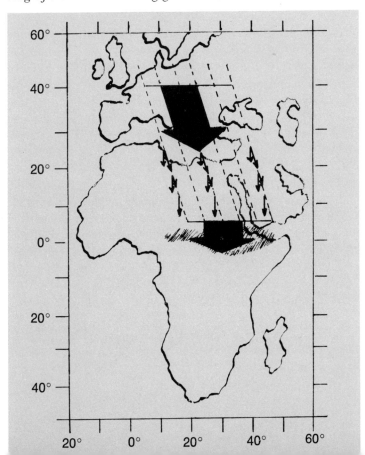

What then of longitude? Human navigators, certainly, have found longitude far harder to determine. The ancient Greeks could find their latitude from the elevations of the sun or the fixed stars, but it was not until 1761 that navigators had a chronometer accurate enough to measure longitude to half a degree.[6] Equipped with an accurate chronometer set to Greenwich Time, a human navigator can find his longitude by comparing his own local time—shown by the time of sunrise, for example—with that of his Greenwich chronometer.

A bird has a very different problem. True, it has an accurate internal clock, but that clock does not maintain a fixed time as the chronometer does; it undergoes continual resetting to local time wherever it happens to be. So how could a bird know the longitude from the positions of the stars?

One way would be through an ability to relate the diurnal rotation of the constellations around the polestar to longitude. This certainly would seem difficult, but the Sauers thought that it should be investigated. In the planetarium they rotated the constellations around the polestar to correspond to the actual aspects of the night sky at various longitudes, which is the same as saying in different time zones, and observed the flight directions of several species of warblers.

The results were strikingly different from the clean-cut effects that they had observed when they changed the latitude shown by the planetarium sky. The longitude changes caused the birds to behave in a way that was very hard to interpret, and in some cases could not even be duplicated. Nevertheless, the Sauers presented the complete results of their work at the Cold Spring Harbor Symposium on Quantitative Biology in June of 1960.[7] It has since been rather generally agreed that although one bird seemed to choose the direction that would

---

[6] Samuel A. Goudsmit and Robert Claiborne: *Time* (New York: Time, Inc.; 1966), p. 81.
[7] E. G. Franz Sauer and Eleanore M. Sauer: "Star Navigation of Nocturnal Migrating Birds," in *Cold Spring Harbor Symposia on Quantitative Biology*, Vol. XXV (1960), pp. 463–73.

have carried it back to its home locality, the data were too scanty and too variable to be convincing.

And that, as far as the literature shows, and as far as I have been able to learn from Dr. Sauer, was the last report on the Zeiss planetarium experiments with warblers.

In 1965 Dr. Klaus Hoffmann of the Max-Planck-Institute for Behavioral Physiology, a noted authority on bird behavior, wrote, "To clarify the situation, and to get further insight into the clock mechanism involved, more diversified experiments under planetarium conditions seem necessary, including experiments with birds whose clocks have been [artificially] shifted. As far as I am aware, no such experiments have been performed."[8]

Until someone takes up the problem again, with the specific aim of learning more about whether birds can actually sense longitude (and if they can, how they use the information for navigation), our understanding will remain about where the Sauers left it in 1960.

As to what may lie ahead, Donald R. Griffin, who is professor of zoology at Harvard, and who has been studying the problems of bird migration and navigation for more than thirty years, has made the following forecast.

"Real advances in understanding a subject like bird migration almost always come as partial or complete surprises. A generation ago neither I nor any other scientist would have predicted with the slightest confidence several of [these] discoveries about bird migration. . . . Future advances may be equally unexpected. If scientific progress were predictable, it would become a sort of engineering, useful perhaps, but much less fun."[9]

[8] Klaus Hoffmann: "Clock-Mechanisms in Celestial Orientation of Animals," in Jürgen Aschoff, ed.: *Circadian Clocks* (Amsterdam: North-Holland; 1965), p. 435.
[9] Donald R. Griffin: *Bird Migration*, p. 171.

I once read a . . . fairy tale,
called The Three Princes of Serendip:
as their highnesses travelled,
they were always making discoveries,
by accidents and sagacity,
of things which they were not
in quest of: for instance
one of them discovered that a mule
blind of the right eye had
travelled the same road lately,
because the grass was eaten
only on the left side,
where it was worse
than on the right—now do you
understand Serendipity?

HORACE WALPOLE,

1754

13. Harker: Cockroaches and Cancer

On a warm June day in 1960 Dr. Janet Harker, Fellow of Girton College, Cambridge University, rose to tell her scientific colleagues about the meticulous experiments on which she had labored for seven years—using the common cockroach. Gathered at the Cold Spring Harbor Laboratory of the Long Island Biological Association were nearly one hundred fifty international experts on biological clocks, including more than thirty from abroad. Men and women like Jürgen Aschoff, Frank A. Brown, Jr., Erwin Bünning, J. L. Cloudsley-Thompson, Franz Halberg, Karl C. Hamner, Martin Lindauer, Syuiti Mori, Colin S. Pittendrigh, Kenneth S. Rawson, Max Renner, Arne Sollberger, Marguerite Webb, and William Wolf had gathered from Germany, the Sudan, Japan, Sweden, England, Italy, and the United States for a ten-day symposium on the rhythms of nature.

In a firm clear voice, modulated with a trace of her native Australian speech, Janet Harker began: "Present-day studies on the problem of circadian rhythms seem to fall into three groups: biochemical and biophysical studies of the controlled mechanism at a cellular level, behavioral studies on the whole organism, and physiological studies on the way in which cellular rhythmicity is translated into the overall behavior of the organism.

"This last approach is beginning to reveal something of the complexity which the endogenous controlling system may attain. Since it might not enter the field of the other two approaches, it is perhaps of particular interest that among the interacting processes we are finding some which, while not in themselves rhythmical, are essential for the continuance of the rhythmical functioning of certain cells.

"The participation of an endocrine system in the locomotor activity rhythm of an insect was first shown in *Periplaneta americana*. Cockroaches taken from continuous light, and having no measurable activity rhythm, were joined . . ."[1]

What Dr. Janet Harker told the experts, in the specialized language of their own discipline, was that for the time being she had stopped worrying about whether living clocks are inherited or are driven by subtle external forces, and had turned to a search for the location of the clock in a single species. If she were successful in pinpointing the exact site to a single cell or to a small group of cells, it should then be possible to study the physiology of these cells in detail, and perhaps to throw new light on their underlying mechanism. The results she summarized in her paper illustrated the very essence of science, for she was able not only to gain deeper insight into the nature of living clocks, but also to present

---

[1] Janet C. Harker: "Endocrine and Nervous Factors in Insect Circadian Rhythms," *Cold Spring Harbor Symposia on Quantitative Biology*, Vol. XXV (1960), pp. 279–87.

the world of science with a tantalizing new puzzle that cried out for further research to solve it.

At the close of her presentation she was amazed by a spontaneous audience reaction seldom seen in the normally staid atmosphere of scientific meetings. Her colleagues sprang to their feet and gave her an ovation—an accolade for her extraordinary laboratory technique, and for the new clue she had found that might one day lead to a better understanding of the causes of cancer.

Janet E. Harker was born and brought up in Australia, and even before she had reached kindergarten age she knew what her life work would be. "It is difficult to answer . . . why I decided to become a biologist," she wrote informally, "since the decision was made when I was about five, anyway before I started school, and I can't imagine what reasons I had, let alone that there was such a subject!"[2] But become a biologist she did, and her brilliant work earned the Royal Zoological Society Gold Medal in 1962.

She did her undergraduate work at the University of Sydney, continued there for her M.Sc., and then went to England to earn her Ph.D. at the University of Manchester. In 1952 she received a faculty appointment in Girton College, Cambridge.

For perspective, let us look at the University of Cambridge as an institution. It is a man's world. It originated in the year 1209 when some Oxford scholars, after an altercation with the townsfolk there, moved to Cambridge. Its history is too illustrious to need detailing here, but Cambridge science and Cambridge scientists have been world-renowned for centuries. Now, in the second half of the twentieth century, it is no mean feat for a woman to gain admission to Cambridge even as a student. Girton College has facilities for only seven hundred students, and young English women are so eager for Cambridge degrees that only the top three percent of qualified

---

[2] Janet C. Harker to Ritchie R. Ward, personal communication, March 28, 1965.

applicants are admitted. For a woman to win a faculty appointment at Cambridge is indeed notable.

At Girton, Janet Harker found herself immediately immersed in a heavy load of teaching and administrative duties, but she had ample energy to take up her research on biological clocks with both hands. Over a very long and very tedious period, she gradually developed the delicate touch that was essential for the minute surgical operations she was to perform on cockroaches—operations that have been the despair and frustration of other workers who have tried to duplicate them. At the same time she broadened and deepened her interest in physiology, in particular the intricate physiological processes that take place in cells.

One might suppose that working at Cambridge Janet Harker had the most modern precision equipment at her command. Actually she had nothing of the sort. Professor Franz Halberg, of the University of Minnesota Medical School has since described the conditions under which she worked as "brutal—as though I had to do pathology with a student microscope." But Janet Harker did not see her situation as all that brutal.

In her own words: "Perhaps the worst difficulty involved in this work was one which is common to most work on rhythms—I had to do these operations at the same time each day, and the whole process took several hours. The only time I was likely to be undisturbed, day in and day out, was very early in the morning before the day's work started, so most of the work was done between about 4:00 and 8:00 A.M.—and of course there could be no days off. Still it was not as bad as other work I have done on rhythms which have kept me with an experiment for as much as 40 hours at a time, or repeated 24 hours at a time at frequent intervals.

"Not much physical equipment is needed for this sort of work. It really depends on very fine dissecting instruments, and tiny chips of razor blade and forceps made from very fine tungsten wire proved better than anything which could be bought. It is true that Halberg was rather shocked to find that I had done hundreds of mitotic counts without the aid of a mechanical stage on the rather simple microscope I then had, but apart from a bit of strain, I do not know that it made a great deal of difference in the end."[3]

This was the kind of effort that Janet Harker was prepared to give to learn the exact site of the clock in a living organism. But where should she start? Why did she choose *Periplaneta americana,* the common cockroach, rather than some other living organism? Many different plants and animals possess readily observable rhythms, but a great many of them are difficult to handle in the laboratory.

Janet Harker chose the cockroach for a number of reasons. In the first place, the cockroach times its activities quite precisely. It is a nocturnal animal, and begins to run about just as darkness falls. Kept in the laboratory with the lights set for exactly 12 hours on and 12 hours off, cockroaches begin their running activity within minutes after it is dark. Activity rises to a peak some two hours later, and then after three or four hours, the animals quiet down and remain quiet for the rest of the dark phase and throughout the following 12-hour light phase. As soon as it is dark once more, they start running again.

What happens then if cockroaches are kept in continuous light for days at a time? Exactly the same sort of thing that de Mairan observed with his sensitive plants nearly two centuries earlier. Cockroaches know when dark *should* fall, and begin running when the lights *should* go out. Then gradually, after being exposed to many days of continuous light, they lose their time sense and begin running at completely random times. Since this running activity serves as the conveniently observable "hands" of its internal clock, the cockroach is an excellent animal to study: it can time itself to precise 24-hour cycles, it continues its cyclic activities for days even in continuous light, and it slowly loses its rhythm if exposed to steady light for long enough. It is then said to be "arhythmic."

[3] *Ibid.*

The cockroach is a good laboratory animal for other reasons also. It lives for months, or even up to a year, in the laboratory; it is large enough to be conveniently dissected; it is very tough (although a few may die from just falling); and it can be fed practically anything. For such reasons, much work has been done with it by biologists, and a large fund of background information about it is available.

Finally, the cockroach is the most primitive of living insects that have wings. Its central nervous system is not well developed, and its various parts have a great deal of autonomy. Thus a decapitated cockroach can live for days. It cannot see, of course, and it cannot eat, and its energy supply gradually runs down. But it can run about, and it can copulate; and many aspects of its headless behavior can be studied.[4] Or, if a cockroach's legs are cut off, the rest of its body will behave much as though nothing had happened. If the experimental design requires it, the cockroach's body can be immobilized, as by coating it with wax or a rigid plastic.

This is the creature in which Dr. Janet Harker made the first discovery of the precise location of the living clock. Let us visit her Cambridge laboratory on the eve of the discovery.

"I knew from the earlier experiments I had done," she explains, "that a hormone was involved in the rhythms of running activity. The task was one of tracking down where this hormone came from. I started, of course, by removing the known endocrine organs and seeing if the clock stopped in their absence, but nothing came of this. So I was faced with the problem of virtually taking the animal apart, eliminating one region after another in different animals—this went on for over a year—until I finally tracked down the source of the hormone to the subesophageal ganglion."

Let us pause for a moment to see what Dr. Harker means by this last term. All insects have two brains. The primary one is above the mouth region; the secondary brain is smaller, but governs the insect's activity more directly. This secondary

---

[4] Professor Roderick Craig has remarked that some higher animals have been known to copulate without using their heads.

brain consists of a cluster of nerve cells called the ganglion, and is found just below the esophagus, or gullet; hence the term "subesophageal ganglion." It is an organ about the size of a pinhead, and accessible to skilled surgery under the microscope.

Dr. Harker continued. "I spent nearly three years cauterizing minute regions of the subesophageal ganglion with a high-frequency cauterizer, which precipitates the protein in just a few cells at a time, until I ultimately found the four neurosecretory cells that are necessary for the maintenance of the rhythm.[5]

"Because it is possible to transplant endocrine organs into the blood stream of other cockroaches, I was able to check that the cells I had found were really responsible for secreting the hormone at precise time intervals; nerve connections were not responsible for their time-keeping."[6]

Now she knew where the cockroach kept his clock. But she did not know what made it run, and she burned with curiosity to find out.

One way to study how a mechanism works is to throw it out of kilter and see what happens. But living clocks are very hard to upset, and until Janet Harker had found the function of the four nerve cells that periodically release hormone, there was no physiological way to confuse such a clock.

"I thought," she added, "that having found this one clock, it might be possible to expose an animal to two clocks running out of time with each other. And that this might have an effect on the animal similar to that which would occur if a clock were malfunctioning."

The entire experiment was an extremely long one, because it meant surgical transplant of neurosecretory cells every day into large numbers of animals. In a masterpiece of understatement Janet Harker said: "It was a rather fiddling task."

She attacked the problem in three steps. The first step

---

[5] Neurosecretory cells are cells that both secrete hormone and transmit nerve impulses.
[6] Harker to Ward, March 28, 1965.

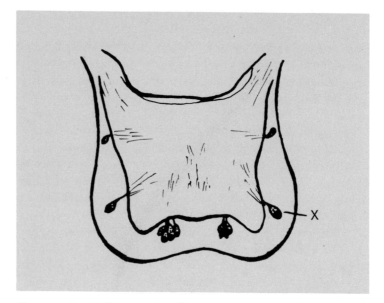

Cross section of the secondary brain of a common cockroach, showing the positions of the four cells (X) that are concerned with rhythmic running activity.

was to find out whether the cockroach's clock could be transplanted from a donor to a host without damage, and if so, whether the host would respond to the time of the new clock. Keep in mind the fact that cockroaches accustomed to the normal day-night cycle begin running just at nightfall, and cockroaches exposed to continuous light lose their time sense. Would it then be possible to restore an arhythmic cockroach's time sense through the clock of a rhythmic one?

To test this, Dr. Harker joined two cockroaches like Siamese twins, one on top of the other. The bottom cockroach was the one that had no rhythm, but was completely free to move. The top one had the clear rhythm of the normal day-night cycle, but Dr. Harker removed its legs and fixed its body rigid so that it could not wriggle. Under these conditions, and with the only communication between the two insects being through their body fluid, the bottom cockroach immediately took up the rhythm of the top one, and began running on schedule. It was clear that the bottom cockroach was responding to time signals transmitted through the bloodstream.

Next, Dr. Harker produced arhythmic cockroaches by surgically removing their clocks. She decapitated specimens, making sure that she also removed the subesophageal ganglia. Then she retimed the bodies by implanting ganglia from normal specimens into their abdomens. She found that the headless bodies followed the rhythms of the donors for days. Thus she showed that the neurosecretory cells could continue their established rhythms after all nervous connections had been broken.

The final step was to throw the cockroach's timing mechanism completely off balance. One further bit of information is needed to understand this group of experiments. If cock-

This experiment showed that cockroach time signals are carried through the bloodstream, and not by nerve connections. A normally timed cockroach, above, was joined to a rhythmless cockroach, below, with no intercommunication possible except through the body fluids. The rhythmless cockroach immediately took up the rhythm of the normal one.

roaches are placed in a light-dark cycle the exact reverse of normal, they eventually learn to start running when the lights are turned off at dawn, and remain quiet when they are turned on in the evening. We may say that although they live in England, they run on New Zealand time.

Remember that Dr. Harker had already narrowed the site of the clock down to the four cells in the ganglion. Since her objective was to expose a single animal to two clocks running completely out of phase with each other, it was a logical step to transplant neurosecretory cells from New Zealand cockroaches into Cambridge ones.

But first, there was the usual requirement of the control experiment. Could she transplant single *cells* from one cockroach to another at all? She could and did, first with animals all timed to the normal day-night cycle, and she found that her surgery did them no harm. The insects continued as lively as ever, on donor clocks timed the same as their own.

Then came the crucial experiment. Dr. Harker took normal Cambridge cockroaches and implanted into them clock cells from cockroaches that were running on New Zealand time. What would happen? Certain cells were telling the confused animal that it was daytime, and others were saying that it was night. Here was internal stress with a vengeance!

Let us hear the story in Janet Harker's own words. "Turning then to the midgut tumor story—it was really the result of the same sort of tedious long-term work, although the outcome was a complete surprise. Because biological clocks are so difficult to upset in any way there had previously been no way of discovering what happened if a clock went wrong. But I thought that having found this one clock it might be possible to expose an animal to two clocks running out of time with each other (by transplant methods), and that this might have an effect on the animal similar to that which would occur if a clock were malfunctioning. The whole experiment was an extremely long one, as it meant transplanting neurosecretory cells daily into animals in large numbers. . . . The outcome, as so often happens, was totally unexpected, and since tumours in insects are very rare it was, looking back, surprising that I ever found out what was happening; it could easily have been overlooked."[7]

What was happening, it appeared, was that the cockroaches were subjected to such severe internal stresses that they quickly developed intestinal cancer and died.

Why should exposure to two clocks running out of phase from each other cause such severe stress in an animal? Why should such stresses lead to cancer? Or do they? Just because one event follows another, it does not necessarily follow that the first causes the second. The problem of establishing the causal relationship is one of the most difficult in all of biology. Learning the full answer to the problem of the mistimed clocks presents even further difficulties, because the next logical step would be to make similar experiments on other kinds of animals. But this cannot yet be done; so far, no single clock has been discovered in any organism other than the cockroach. Still it is encouraging to know that by 1965 Janet Harker thought that several workers in the United States might be close to a breakthrough in mammals.

In reviewing Dr. Harker's work, J. L. Cloudsley-Thompson has said: "[Her] observation that tumors of the mid-gut develop in normal cockroaches when subesophageal ganglia from insects kept under reversed lighting regimes are implanted in them may prove to be of great importance in the study of diseases resulting from stress syndromes in man."[8]

At about the time that Cloudsley-Thompson wrote, military medical men in Washington were advising the Joint Chiefs of Staff against requiring ranking generals and admirals to report on vital matters immediately after long jet flights. The doctors had found that high-speed flights across many time zones severely upset the internal clocks of men. The human mind does not work efficiently without time to

[7] *Ibid.*
[8] J. L. Cloudsley-Thompson: *Rhythmic Activity in Animal Physiology and Behaviour* (New York: Academic Press, Inc.; 1961), p. 171.

recover from such stresses.[9] But these are relatively mild time-stress effects. A man can jet from Tokyo to London and recover all of his regular bodily functions within two weeks. What then of the time stresses that may result from space flight? This critical problem of today's technological world will have our full attention in later chapters.

To return to Dr. Janet Harker, the standing ovation that she received at Cold Spring Harbor was not the last reaction she was to hear from other scientists about her work. As would surely follow, other biologists in other parts of the world attempted to duplicate her results. Some of them failed completely, and began to question the reliability of her work. Two chief criticisms arose. The first was that her results could not be substantiated by others, and the second—rather vague—was that the work was not quantitative. Dr. Harker herself made no direct reply to these criticisms beyond adding further experimental proof that she could duplicate her own reported effects over and over again.

One of her colleagues at Cambridge, Dr. E. A. C. Mac-Robbie, who had been intimately familiar with her work for years, has been kind enough to give her own views on Dr. Harker's research: "You mention criticism of her work— repeatability and quantitative nature. On the first point my reaction is that the work has now been repeated independently in Cambridge. The techniques used in some of the earlier attempts to repeat it by other workers were rather crude. The operations are exceedingly delicate, and require a fine hand and considerable practice. It is no use trying a few animals and deciding it doesn't work—persistent attempts in large numbers are needed to master the skill, although once the technique is achieved most operations are successful and the animal survives. In at least one series of attempts to repeat the work that I know of, the presence of the salivary duct very close to the subesophageal ganglion was simply ignored; when this is cut to expose the subesophageal ganglion,

[9] Samuel W. Bryant: "What Jet Travel Does to Your Metabolic Clock," *Fortune*, Vol. LXVIII (November 1963), pp. 160-8, 183-6.

very powerful enzymes are released and damage the cells of the ganglion. It is not surprising that these experiments gave negative results.

"I'm afraid I do not see in what respect the work is not quantitative; I fail to understand this criticism. By its very nature an implant experiment gives a 'yes' or 'no' answer; it is not a statistical effect but one in which a positive result has a definite meaning."[1]

More time will surely pass before biologists reach full agreement on this phase of Dr. Janet Harker's work, but if we sense the general history of science in the right light, it seems destined to hold a high place.

Meanwhile Dr. Harker continues with her administrative duties, her teaching, and her research at Cambridge, and as she accumulates new results that she feels are worth reporting, she publishes them in the scientific literature and presents them at meetings of scientific societies. An ebullient English colleague of hers, who has since accepted a faculty post at one of America's leading technical institutes, has given a picture of her as one who is "very feminine, dresses smartly, and is quite a handsome woman. She is so attractive that at the end of a day's scientific symposium, she receives one invitation for drinks and another for dinner. She gets around!"

[1] E. A. C. MacRobbie to Ritchie R. Ward, personal communication, May 12, 1965.

Natura non facit saltus.

1751

14.   Sweeney: The One-Celled Clock

In 1751 the pioneer botanist Carolus Linnaeus could surely say with confidence that "nature does not proceed by leaps." In his day, in fact, very nearly the same thing might be said of man's understanding of nature. Before the middle of the eighteenth century science had not taken many great leaps forward.

With what wonder, then—with what total disbelief, indeed—might Linnaeus have faced the prospect that a mere two hundred years later another botanist would determine the activity of one plant by measuring the volume of a gas that it gave off in *thousandths of a millionth of a quart!*

Yet Professor Beatrice M. Sweeney did just that, in the late 1950's, at the Scripps Institution of Oceanography. And, apart from the fact that she suffered a great deal of eyestrain, and spent long continuous hours in the laboratory for month after month, she does not seem to regard the achievement as much more than a present-day scientist's normal research activity. Her previous graduate training under Kenneth V. Thimann at Harvard, and her extensive collaboration with J. Woodland Hastings on the luminescence of marine organisms, had prepared her for it.

But for what possible purpose did Beatrice Sweeney want to measure a few thousandths of a millionth of a quart of oxygen given off in an hour by a single-celled plant?

In its simplest terms, she wanted to find out whether that single cell would show the same rhythmic behavior as a massed population of a very large number of the plants. This would be one phase of a long-range program aimed at gaining deeper insight into the internal workings of living clocks.

In a certain sense, a modern biologist is very much like a highly skilled mechanic. With a wide range of tools at his disposal, he selects the one best suited for the work at hand. By contrast, the earlier naturalists studied a particular plant or animal simply because it was interesting. For example, Charles Darwin made a lengthy study of carnivorous plants just because he was fascinated by the ways in which they trapped and ingested insects. A biologist today, however, would be more apt to have a particular problem or theory in mind, and would choose the plant or animal that he thought might be the most useful tool to use to study that problem; he would choose that most promising tool with no special regard to its interest in any other respect.

Sweeney's choice was influenced by several primary considerations. In the first place, as a botanist, she would naturally choose a plant to work with. In the second place, she was interested in studying the physiology of living clocks, and she therefore wanted a plant with clearly defined displays of its physiological behavior with time—that is, she wanted an especially clear view of the hands of the living clock. She also wanted an organism whose physiological responses depended on biochemical reactions that were reasonably well

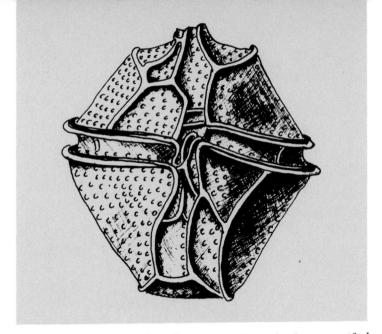

Exterior view of a single cell of Gonyaulax polyedra, *magnified 2,000 times. As an aid to distinguishing this species from the many other dino-flagellates, the two hairlike flagella that the cell possesses have been omitted from the drawing. Normally one of these is found wrapped around the "girdle" at the wide part of the cell, and the other trails from its base.*

understood. And of course for convenience, she wanted an organism that was readily available and easy to culture in the laboratory.

Thus it was no simple tool that she needed, but she was fortunate in that she had an excellent one ready to hand. She and Hastings had already used it for a number of other purposes, and she thought she could adapt it to her present needs.[1] This was the single-celled green plant *Gonyaulax polyedra*, which is a marine dinoflagellate. It multiplies rapidly by cell division in sea water or artificial culture medium and is the organism which is responsible for the lighting up of

[1] J. Woodland Hastings and Beatrice M. Sweeney: "On the Mechanisms of Temperature Independence in a Biological Clock," *Proceedings of the National Academy of Sciences, Washington, D.C.,* Vol. XLIII (1957), pp. 804–11.

the "phosphorescent sea." The cell is some two thousandths of an inch in diameter—barely visible to the naked eye— and under a low-power microscope, a single plant cell can be easily picked up out of a culture on the point of a needle.

But while an individual *Gonyaulax* plant consists of but a single cell, its display of rhythms is more complex than those of many higher animals. The most prominent of these displays is that of luminescence. The cells light up under various circumstances with a blue-green light, but especially when they are disturbed. Thus when sea waves teeming with *Gonyaulax* roll at night, a shimmering glow of bluish light follows the crest of the wave, and then slowly dims and dies as the subsiding wave laps the shore.

The same display may be readily seen in the laboratory. If a culture of *Gonyaulax* cells in a tube of sea water is shaken by hand, the water flashes. Or if one wants to control the disturbance more precisely, he may bubble air through a culture tube, and even (by means of a photomultiplier tube) measure the amount of light emitted.

The observer may also see rhythms of persistent flashing in cultures that are left completely undisturbed. This may be independent of the primary disturbance rhythm, or, as Cloudsley-Thompson has suggested, it may be related to it: individual cells certainly bump into each other as they move about in the culture, and also into the walls of the container, and that might well produce an effect similar to shaking.[2]

Less obvious than these two rhythms of flashing is the very faint but persistent emission of light by *Gonyaulax* cells sometimes called "glow"; this glow, faint as it is, has also been found to be rhythmic. Finally, there is the regular rhythm of cell division, by which the plant multiplies itself. Thus Sweeney's requirement for a clock with readily observable hands was apparently satisfied.

As to the requirement that the clock itself depend on known biochemical reactions, that was provided by the lu-

[2] J. L. Cloudsley-Thompson: *Rhythmic Activity in Animal Physiology and Behaviour* (New York: Academic Press Inc., 1961), p. 194.

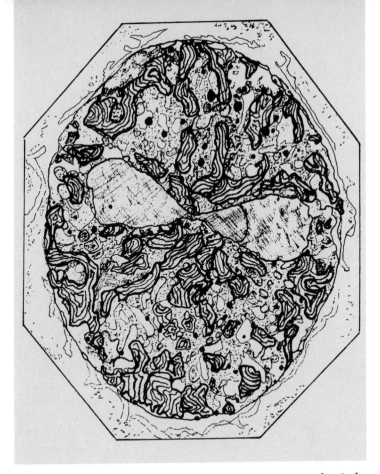

Drawing after a G. Benjamin Bouck electron micrograph of the cross section of a Gonyaulax *cell. Somewhere in the nucleus, mitochondria, chloroplasts, or other structures the clock may someday be found.*

minescence. Such luminescence depends on enzyme-controlled reactions in which the system that emits the light consists of two components: the enzyme itself (in this case called luciferase), and the substrate, or substance on which the enzyme reacts (called luciferin). We should immediately point out that although volumes have been written on bioluminescence, the exact nature of the luciferase and the luciferin in *Gonyaulax* is not very well understood.

This, then, was the basic tool that Sweeney had at hand.

But she could use none of these clock hands that she had studied with Hastings to attack the particular problem that she was now most curious about.

Why? Because she now wanted to study the behavior of a living clock that had been exposed for some time to continuous bright light. Under such continuous bright light, the luminescence of *Gonyaulax* died out. Those hands of the living clock did not work.

Let us return to her reasons for wanting to study the behavior of a single-celled organism in the first place. This stemmed from a very fundamental question about the behavior of all living clocks. It had been known for a long time that when plants and animals that had been accustomed to normal conditions were placed under continuous bright light (for twenty-four hours a day), the normal rhythm gradually died out. For example, Erwin Bünning had shown this effect with the sleep movements of bean leaves as early as 1931. The leaves of a normal bean plant, accustomed to rising and falling day and night, simply stopped moving when placed under continuous bright light. Precise laboratory observations of the rhythms of other organisms had shown the same effect. So also had Sweeney's own cultures containing thousands of *Gonyaulax* cells in a single tube.

What could cause such behavior?

Of course this question involves many related ones, but the primary question that had to be answered first of all was simply this: when a group of cells loses its rhythm under continuous bright light, does every cell in the group lose its rhythm, or does each cell retain its own individual rhythm, but drifting so far out of phase from the rhythms of other cells that the rhythm of the group as a whole would *appear* to have lost its rhythm? Do the individual rhythms drift so far apart that their effects cancel each other out?

The most direct way to answer that question was to find out whether a *single* cell would lose its rhythm under continuous bright light.

What possible experimental methods could Sweeney use

to answer that? What rhythm might continue that she could observe under continuous bright light? She thought of one possibility, and she tried it, and it seemed to offer possibilities. This was based on the process of photosynthesis, which *Gonyaulax,* even though it has only one cell, carries out like all other green plants.

By 1785 the great French chemist, Antoine Lavoisier, had shown that the basal metabolic process in animals consists of the slow burning of waste matter from the lungs by oxygen from the air, yielding carbon dioxide as the combustion product. At about the same time, Joseph Priestley in England had discovered that plants have the power to change the composition of the air that surrounds them. Plant physiologists later found that plants have two metabolic processes:

First, the same slow combustion as that of animals, in which oxygen is consumed and carbon dioxide produced, and

Second, the photosynthetic process, by which the energy of the sun is used, in the reverse process, to consume carbon dioxide from the air and release oxygen to it.

Ordinarily the rate of photosynthesis in a green plant is some ten to thirty times greater than the rate of combustion, so that the effect of the combustion is completely masked. It is as though nothing but photosynthesis were going on, and the green plant was simply giving off oxygen.

Now it had already been shown that cultures of many cells of *Gonyaulax* have a regular daily rhythm of photosynthesis, and further, that this could be followed in the laboratory by measuring the volume of oxygen that the culture produced at any given time during the day. Even this was no simple task, for the amount of oxygen given off even by thousands of single cells is very small, and to measure it required highly refined microchemical techniques.

Now, however, Beatrice Sweeney set herself the task of measuring the volume of oxygen produced by just one of these cells!

For this, no ordinary technique—if the refined techniques of microchemistry can be called ordinary—would come even close to meeting her requirement. She needed a method that would measure volumes of oxygen in a few thousands of a millionth of a quart.

Whenever a problem of such difficulty arises, it is often useful to break completely away from tradition—to put out of mind everything that has been done before—and to search for something entirely novel.

What could be more novel than to look for the answer to such a sophisticated laboratory problem among children's toys?

Everyone is familiar with the Cartesian diver. It was discovered by René Descartes in the seventeenth century, and depends on the principle that a floating object displaces its own weight of water. In its usual form, the diver itself is a glass tube with a small bulb at the top; the tube is filled with just enough water so that the diver will barely float in a glass of water. A rubber diaphragm is stretched across the top of the glass, and when a child presses the diaphragm down with his finger, the air in the bulb is compressed, a little more water enters the diver, and it sinks. When the child releases the finger, the air expands again and the diver rises.

By a sophisticated refinement of this simple principle, P. F. Scholander and others had already been able to measure the oxygen consumed by marine eggs and polychaete worms, and now by further refinement, Sweeney adapted the method for measuring the oxygen production of a single *Gonyaulax* cell.[3]

Instead of the glass diver of the child's toy, Sweeney used a short length of fine capillary saran plastic tubing, drawn out even finer after heating in a flame. One end was sealed off and the other end left open, so that a little air remained trapped in the tiny capillary hole in the center of the tubing. The length of the plastic diver was cut precisely, so that at normal pressure the diver would stay just barely

---

[3] P. F. Scholander, C. L. Cliff, and S. L. Sveinnson: "Respiratory Studies of Single Cells," *Biological Bulletin*, Vol. CII (1952), Part I, "Methods," pp. 157–77.

The principle of the Cartesian diver can be used to measure volumes of oxygen as small as a thousandth of a millionth of a quart.

resting on the bottom of a tube of culture solution. If the pressure over the solution was lowered slightly, then the air trapped in the capillary would expand to form a tiny bubble at the end of the tube. As the size of the bubble grew with further reduction of pressure, the diver would lift off the bottom and slowly rise. By fine adjustment of the gas pressure over the culture solution, Sweeney found that she could keep the diver exactly balanced at a fixed height.

Her method of measuring the oxygen output of a single cell was then straightforward. When she placed such a single cell in a small glass tube together with the diver, the oxygen given off by the cell would raise the gas pressure over the solution, shrink the bubble at the end of the diver, and cause the diver to sink. But by lowering the pressure by a supplementary control system, she could offset the increase in oxy-

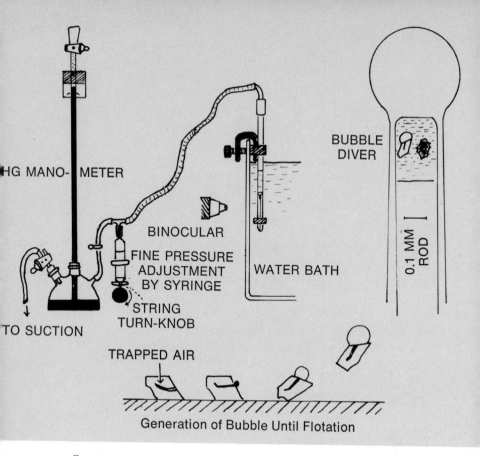

HG MANO- METER

BINOCULAR

FINE PRESSURE
ADJUSTMENT
BY SYRINGE

STRING
TURN-KNOB

TO SUCTION

TRAPPED AIR

WATER BATH

BUBBLE
DIVER

0.1 MM
ROD

Generation of Bubble Until Flotation

Respirometer for measuring oxygen output of a single Gonyaulax
cell. UPPER LEFT, *the complete apparatus, showing the pressure
adjustment and measuring system at the left, the binocular microscope trained on the Cartesian diver, and the tube containing the
diver and the cell in the constant temperature bath at the right.*
UPPER RIGHT, *a fine glass tube (greatly enlarged) holds culture
solution containing the diver and a single cell.* BELOW, *how the
diver rises as the pressure is lowered; air is trapped in the capillary
and expands, forming a bubble that floats the diver.*

gen pressure and keep the diver at its former height. Thus the
change in pressure, which she measured by the height of a
column of mercury, served as a measure of the amount of
oxygen produced by the cell.

The unbelievable precision of the entire operation can be

further appreciated from the fact that Sweeney had to keep her pressure adjusted to such fineness that the diver kept its height level with the hairline of a microscope that she had trained on it. More, since all gases expand and contract as the temperature rises and falls, she had to keep the tube containing the diver and the cell in a constant-temperature bath that did not vary more than one-twentieth of a degree Fahrenheit.

Now with her complete respirometer apparatus and technique in hand, Sweeney could proceed to explore her question as to whether the rhythms of single cells were the same as those of a population of them.

She had an ample supply of *Gonyaulax* cells that she kept in a nutrient solution under conditions of alternating light and darkness: 12 hours of light and 12 hours of darkness. She expected that when she transferred one of these cells to her respirometer, its photosynthetic activity would rise shortly after dawn to some maximum around midday, and then fall with the approach of twilight.[4] Her experiments amply confirmed this, but the proof was not easy, for she encountered an unexpected complication. This was that no single cell was capable of continuing its photosynthetic activity for more than 14 hours.

Sweeney herself was quite willing to keep her eye glued to the microscope eyepiece, continuously adjusting the pressure in the respirometer so that the diver held level with the microscope hairline. In fact, she even tried to extend the lives of the cells with penicillin, but this did not succeed. Thus it was the plant cells that tired out to end an experiment—not Sweeney!

However, she found a way around that difficulty also. If a cell stopped functioning after 14 hours, she could start

[4] Ideally, she would have maintained a cell in the respirometer under exactly the same conditions as it had in the original culture, that is, 12 hours of light and 12 hours of darkness. But of course she could not see the diver in the dark, and she worked out a method for correcting for the extended light period by means of an appropriate calibration.

one cell at dawn and observe it for the 14-hour period. She could then begin a following experiment an hour after dawn, and continue that for the next 14 hours. A third experiment could begin 2 hours after dawn, and so on, until she had extended the time span covered by all of her experiments from dawn to 25 hours after dawn.

She could then plot the individual points from all of these observations to give a composite curve. This she did, and it turned out to display the typical diurnal rhythm pattern that she expected. At dawn, the amount of oxygen a single cell gave off was about 2 thousandths of a millionth of a quart per hour (TMQ/H). Five hours later it had risen to nearly 4, or twice the initial value; 14 hours later it sank back to 2. From that minimum, it began rising again, as it had from hour zero after dawn. At the end of the twenty-fifth hour after dawn, it had risen back to nearly 4 TMQ/H. The regular diurnal periodicity of *Gonyaulax* single cells was thus clearly confirmed.

Now came by far the most critical experiment. Sweeney prepared a new culture of cells in a nutrient solution, which she kept under continuous bright light. She then placed single cells from that culture in the respirometer and returned to her continuous tedious observation of the level of the diver against the microscope hairline.

How would the rate of photosynthesis now proceed? Would each cell behave differently, so that the net apparent effect of a population of them would seem to be without any rhythm? Would random distributions of many different rhythms all cancel each other out?

Or would every cell be nonrhythmic?

Sweeney's volume of observations made it convincingly clear that all of the cells that had been exposed to continuous bright light lacked rhythm. Their rate of oxygen production at dawn was the same 2 TMQ/H as had been the case with the cells conditioned to the light-dark regime, but the rate never rose after that. It simply stayed at the same level for 16 hours or more. There was no wave form to be seen. No

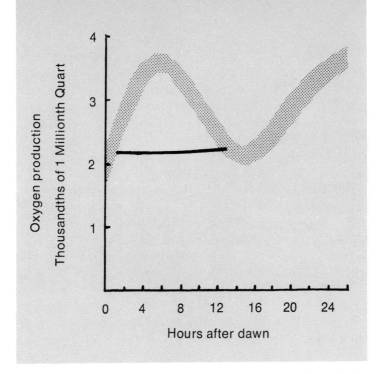

Photosynthetic rhythm of a single Gonyaulax *cell. The shaded cyclic curve is the composite of many separate rhythm measurements of single cells that had been kept in alternating light and darkness. The thinner solid line shows the response of cells that had been exposed to continuous bright light; the photosynthesis was then not rhythmic.*

cell cultured under continuous bright light had any rhythm whatever.

And Beatrice Sweeney's crucial experiments had answered her primary question. A population of cells maintained in continuous bright light loses its rhythm because every cell in the population does so—not because rhythmic cells drift out of phase from each other.[5]

[5] Beatrice M. Sweeney: "The Photosynthetic Rhythm in Single Cells of *Gonyaulax Polyedra,*" in *Cold Spring Harbor Symposia on Quantitative Biology,* Vol. XXV (1960), p. 145. See also Beatrice M. Sweeney: *Rhythmic Phenomena in Plants* (New York: Academic Press, Inc.; 1969), *passim.*

In her laboratory at the University of California, Santa Barbara, Beatrice M. Sweeney studies the rhythms of the one-celled plant Gonyaulax polyedra. She prepares to examine a single cell in sea water on the stage of a microscope, UPPER LEFT, observes an oscilloscope record of the effects of very brief, very intense light flashes, UPPER RIGHT, and notes the progress of a recording instrument chart, LOWER RIGHT. She measured the oxygen output of Gonyaulax single cells in photosynthesis while she was a research associate at Scripps Institution of Oceanography, LOWER LEFT.

Just what kind of woman would be willing to subject herself to the strain on eyes, nerves, and muscles to maintain continuous fourteen-hour watches, over and over and over again, on the position of a tiny Cartesian diver in relation to the reference hairline of a microscope?

A quiet, nerveless, placid individual? Single perhaps?

It would be hard to suggest a less accurate characterization of Beatrice Sweeney. She is a tremendously vital person, with unbounded energy. She moves rapidly about the laboratory, making observations and adjustments to apparatus as though that particular step were the most important yet. She speaks with an incisive New England inflection, and so rapidly that her mind seems to be racing faster than the words can flow. When she lectures, she turns abruptly from blackboard to notes to slides to audience, her thoughts again outpacing her delivery, her pointer moving restlessly. If she is tired at the end of such a lecture, she does not show it, but the minds of some of her slower listeners seem to.

And while maintaining such a professional pace, she has married and brought up four children.

Asked how she could keep up with family responsibilities and pursue such a demanding career, she says simply that she did it because she decided very early that that was what she *would* do. But she is quick to add that no woman could even have begun such a program without the sympathy and understanding of a man like her graduate adviser at Harvard, Dr. Kenneth V. Thimann. When she went to Cambridge, Thimann was a young instructor who had just joined the Harvard faculty, and he says that after he had had a few preliminary discussions with this "tense little girl who had just graduated from Smith College with honors in biology," he accepted her as a graduate student. After she had earned her Ph.D. he continued to follow her career as a "passionate student of nature, [for] nothing could stop her."[6]

For a number of years Beatrice Sweeney was a research

---

[6] Kenneth V. Thimann to Ritchie R. Ward, personal communication, March 14, 1970.

associate at the Scripps Institution of Oceanography (now a part of the University of California at San Diego), and it was at Scripps that she did her work on the photosynthetic rhythms of *Gonyaulax* single cells. She is now associate professor of biology at the University of California at Santa Barbara, and in her research there she is delving deeper and deeper into the physiology of *Gonyaulax* rhythm reactions.

She feels that her laboratory is "making some progress on the composition of the pigment that is associated with the behavior of *Gonyaulax* under constant light," but with characteristic impatience she says that the problem is "very exasperating, because I am not a biochemist."

Still, the exasperation is only superficial, for more seriously she says: "The fun of science is that progress always generates new problems."

15.   Brown: Mysterious Subtle
Timers

Frank A. Brown, Jr., Morrison Professor of Biology at Northwestern University, is undoubtedly the most controversial
figure in the field of biological rhythms today.

Why?

Because he holds very strong views that sharply challenge
the ideas of most workers in the field, and because he is a
blunt and forthright man who says his say without regard
to the outcome. Newcomers, attending their first conference
on living clocks, have been astounded at the violent inter-
changes between Brown and those he singles out as his
"opponents," and have gone home wondering about the
propriety of name-calling at scientific meetings in the twentieth century.

Frank Brown's introduction to biological clocks was so

different from the experiences of other "clock-watchers" that it is hardly surprising that after twenty-five years of fiercely independent research he should have arrived at a position poles apart from that of most biologists.

Soon after Brown received his Ph.D. in zoology from Harvard he worked for a time at the biological Research Station in Bermuda. There he saw at first hand two remarkable examples of biological rhythms that were to influence the whole course of his career. The conspicuous feature about these rhythms was that they were timed to the phases of the moon.

Thus their very nature was inherently different from the daily, or approximately daily, rhythms that had seized the interest of so many other researchers.

One of the lunar rhythms that Brown had found fascinating was the swarming of schools of Bermuda shrimp which occurs at regular monthly periods. Another was the swarming of the Atlantic fireworm—and he still enjoys telling about that:[1]

The late Professor E. L. Mark, of Harvard University, who devoted a great deal of his energy and enthusiasm to stimulating the interests of biologists at the Bermuda Biological Station, took an almost childlike delight in introducing novices to the weird and wonderful phenomena found in the laboratory.

Perhaps his favorite pastime was to await the day when he knew the fireworms were scheduled to swarm. He would then inform newcomers of this spectacular event and tell them to join him at the water's edge at a sharply prescribed time that evening. There he would solemnly perform a "count-down."

"Fifteen minutes more, ten, five . . . Now keep your eyes on the water," he would say. First a few brilliant flashes, then more and more until they seemed to be everywhere. These

[1] This is the same fireworm that Christopher Columbus may have mistaken for a flickering light ashore at his first landfall in the New World (Chapter 2).

were the female fireworms signaling the waiting males that they were ready to discharge their eggs into the water.

The swarming of the fireworms is a case of a critical biological event in a population being beautifully timed by a calendar-clock system to enable the synchronization of all the individuals for the survival of the species. The spawning is pinpointed to a time of year, phase of moon, and finally to a specific time of day.

By this means, nature assures that reproduction will be an orderly, regulated process, with the eggs and sperm present in abundance at the same time and place.[2]

Thus early in his career Frank Brown became intensely interested in the effect of lunar cycles on physiological events.

"But at that time," he says, "I didn't even know that others just at the turn of the century had discovered lunar rhythms. But they had done it, and written it up, and there it was in the literature. But ever since then, people had simply ignored it. It was just not time for it. Anyone who brought in the moon was just a lunatic. And no respectable scientist mentioned the moon. So I stayed out of it for a while until I felt I didn't have anything really to lose."[3]

For several years after his tour in Bermuda, Frank Brown's attention was concentrated on the study of the endocrine systems in crustaceans. But the puzzle of biological rhythms kept nagging at the back of his mind, and he had trouble keeping his interest high in preparing endocrine extracts in the laboratory.

"After all, these were just soups," he says. "What is the meaning of an endocrine in a crustacean anyway? I was finding endocrine systems in crabs and lobsters and shrimps by

[2] Frank A. Brown, Jr.: "Biological Clocks," *Oceanology International* (July-Aug., 1967).
[3] Unless otherwise identified, the quotations in this chapter are taken from notes made during a discussion between Professor Brown and myself on May 26, 1967.

grinding up tissues, but after all, endocrines are really characteristic of vertebrates. Where's the thyroid gland in that shrimp? Where's the pituitary in that lobster? Sure, they have a parallel or analogous system that is coordinating their bodies, but that really doesn't have the interest of the endocrine system of an animal with a backbone. So I got out of the field. And I was glad I did. The literature of endocrinology is enormous, and I don't particularly like to read. I'd far rather do experiments."

The very first experiment that he wanted to do was to find out for sure whether animal rhythms are influenced by temperature—a question typical of the independent thinking so manifest in all his work.

"I got into the field because I simply didn't believe some of the results that other people were finding. There were these two pioneers in the field, Professor John Welsh at Harvard, and Professor Orlando Park, who was then at the University of Illinois, and they were bringing crustaceans and insects into the laboratory and putting them in some arbitrary constant temperature, and they were finding that the rhythms persisted. Now they were bringing them in from cold fields, and also on hot days, and they thought that they were finding that the same rhythms persisted at laboratory temperatures, and I didn't believe it.

"After all, I was brought up under Crozier in general physiology at Harvard, and one of the things that he impressed on us graduate students as most important was this van't Hoff rule that when you warm things up, you speed 'em up, and when you cool them down, you slow 'em down. And so I said, these rhythms can't be the same at different temperatures, because that is not what we were taught."

So Frank Brown's curiosity became thoroughly aroused. He could see a broad field for fascinating research ahead, and then and there he decided that he would make the study of rhythms his career.

Friends warned him earnestly against it. "This is not a field to go any further in," they said. "If you want to make a . . . if you want a successful career, go into something that's likely to provide some sort of an answer in a reasonable time."

⁻Even that did not deter him, and one day soon after the end of World War II, he asked one of his graduate students to come in and talk over a new research project. This was H. Marguerite Webb, whose name has since been linked with Brown's as coauthor of many scientific papers, and is now professor of biological science at Goucher College.

Brown told of the things that Welsh and Park were finding, and he said to Miss Webb: "Look, we've simply got to check this. It's a critical thing. Do these rhythms vary with temperature as van't Hoff's rule says, or don't they?"

And with that, Brown and Webb began the research that was to lead to the first of dozens of papers on biological clocks that were to come from the biological laboratories at Northwestern.

The animal that they chose for their experiments was the little fiddler crab *Uca*, which is seen scurrying around beaches everywhere by the thousands, and sounding—to Rachel Carson's ear—like the crackling of paper. Under normal conditions *Uca* displays a regular rhythmic change in the color of its skin. During the day it assumes a dark color, and at night it is pale; and the degree of color change can be measured in the laboratory using standard procedures (staging of the chromatophores).

Brown and Webb started with the hypothesis that the fiddler crabs would obey van't Hoff's rule, and if they did, the rhythm of color change would double or triple when the temperature was raised by 18°F., and cut to a half or a third if the temperature were lowered by the same amount.

In their first experiments they checked the rhythms of a large population of crabs over an extended period of time at 61°F. Starting with crabs that had been exposed to the normal day-night cycle, they then placed them in continuous darkness at the selected constant temperature. As they had expected, they found that the rhythms continued in constant darkness, but they were amazed at the accuracy with which the crabs followed an *exact* 24-hour period.

"Just as Mendel was very lucky to have chosen the pea as his experimental material for studying hybridization, we were very lucky to have chosen the fiddler crab," says Brown. "It's rhythm is not *circa*dian—about 24 hours; it is *dian*—it is *exactly* 24 hours."

Now Brown and Webb raised the temperature of their crabs from 61°F. to 79°F. Would the rate of color change speed up according to the van't Hoff rule? With increasing consternation, as they recorded measurement after measurement, they failed to find any change in the period whatever. It held to a precise 24-hour cycle. To be sure, the depth of color change that the crabs displayed was higher at the higher temperatures, but the rhythm remained the same.

Next they lowered the temperature from 61°F. to 43°F., and there too they found an accurate 24-hour rhythm persisting.

Recognizing that they might have small errors in measuring the rhythms of color change on single days, they correctly reasoned that even small differences from 24 hours on single days would add up to larger differences over many days. But even by that test, they could find no measureable drift away from exactly 24 hours.[4]

This seemed extraordinary. How could it be explained? Frank Brown was not ready to abandon van't Hoff's rule, and he certainly was not ready to believe that all of the data that they had collected with such care could be wrong either. Yet the crabs seemed to be deviating completely from the accepted laws of physiology.

How could this be possible? No known physiological change within the crab could explain it; and that being so, how could the crabs be timed internally? And if not internally, then how? Certainly not by light and temperature fluctuations, for they had been held constant. Some other external rhythm then? But what? Something as yet unknown or un-

[4] Frank A. Brown, Jr. and H. Marguerite Webb: "Temperature Relations of Endogenous Daily Rhythmicity in the Fiddler Crab *Uca*," *Physiological Zoology*, Vol. XXI (1948), pp. 371–81.

recognized? All right. Then Frank Brown would search for it until he ran it down.[5]

At just that point his path deviated sharply from the main highway that was to be followed by most researchers on living clocks; they felt that the weight of evidence favored an internal clock, and that the most rewarding results would come from learning as much as possible about just how such a clock would function. Meanwhile Brown went wherever his experiments and his imagination led him. Unavoidably that also led him into a battle of ideas with many other biologists that has raged ever since.

"If you decide to go your own way," says Brown, "you must realize from the start that it's going to be a very, very rough road. But one can't really complain about this sort of thing. People who go into new and strange fields just have to be prepared—they simply have to acknowledge that they are going to be subjected to ridicule as part of the cost of the thing.

"But if you want the fun, the excitement, the thrill of making your own discoveries, then you've got to be ready to be sniped at, and all sorts of things both ethical and unethical."

Frank Brown did not have to wait long after the first paper on fiddler crabs was published to get sniped at. Critics suggested that the precise 24-hour rhythms that he had found might not be real—might actually be caused by the very weak light that he and Webb had used while they were measuring the color changes. But Brown had anticipated just that criticism, and had already performed control experiments to show that there was no basis for it. He left control crabs sealed under the same experimental conditions for periods of up to two weeks, and showed that they were fully syn-

---

[5] Brown knew that before the turn of the century the Swedish chemist Svante Arrhenius had suggested that cosmic forces affect physiological behavior (Chapter 3), but he wanted now to test that speculation with a body of solid experimental evidence that Arrhenius would never have had the patience to assemble.

chronized with the experimental groups. "The staging routine was *not* responsible for either the temperature-independence or for the precision of the 24-hour periods."

He chuckles as he tells of that criticism. "Actually, I am a very cautious person—probably one of the most cautious, one of the most conservative people in science today. Working as we do in strange fields, we never publish anything until we have two or three times as much ammunition as we need."

In addition to the amazingly accurate 24-hour rhythms of color change in fiddler crabs, Brown and Webb also found an entirely different rhythm in the crabs that was timed with the moon rather than with the sun. At low tide the crabs run down the beach and scour it for food, and at high tide they run up the beach and rest. Except for some local variations, the tides are in phase with the lunar day of 24 hours and 50 minutes, and thus under normal conditions the fiddlers keep in step with the moon.

Again, would they do the same thing in constant conditions in the laboratory? Again, under carefully controlled conditions, Brown and Webb assembled a large volume of data that showed that they did.

As we have seen so often, one question answered in science immediately asks others, and Brown's next question was: What would happen if one took a tidal creature known to be timed to the moon, and transported it away from its home beach to a distant location where tides (if they could occur there) would be timed differently? Would the animal continue the rhythm of the home beach, or would it shift?

Brown had ready access to almost any tidal creature that he might need. During the summer he usually left Evanston and carried on his research at the Woods Hole Marine Biological Laboratory, near Falmouth on Cape Cod. There he served both as a consultant and a trustee. And thinking of the various tidal animals that were available in the vicinity, it occurred to him that there was one that had a definite tidal rhythm, but that did not scurry around like the fiddler crabs, because

Oysters collected in New Haven, Connecticut, were shipped to Evanston, Illinois, and kept there in pans of sea water in a dark room. At first they continued to open their shells widest when it was high tide in New Haven. By the end of two weeks they had reset their rhythms so that they opened widest when high tide would have occurred at Evanston if it were a coastal city.

it had no feet. It should be more convenient to observe. That was the oyster.

It turned out to be an excellent animal to give him his answer, which he summed up in a few words.

"Oysters collected in New Haven harbor, Connecticut, were shipped to Evanston, Illinois. Kept in pans of sea water in a dark room, they continued for a time to open their shells widest when it was high tide in New Haven waters. By the end of two weeks the New England oysters had reset their rhythms to open their shells widest at the moon's zenith and nadir positions with reference to Evanston. The oysters maintained thereafter this new schedule throughout the month they were observed. The zenith and nadir positions of the moon—the two lunar positions on opposite sides of the earth which give rise to twice-daily high tides—are periods of the moon's maximum gravitational effects upon the atmospheric tides over Evanston; this same gravitational effect would produce high ocean tides if Evanston were a coastal city."[6]

[6] Frank A. Brown, Jr.: "Life's Mysterious Clocks," in Richard Thruelson and John Kobler, eds.: *Adventures of the Mind* (New York: Random House, Inc.: Vintage Books; 1962), p. 159.

But again, Frank Brown regarded these findings simply as presenting a host of new questions. First among these were: (1) are the assumed-constant conditions under which we study our organisms really constant? and (2) if they are not—if they are influenced by some unknown force or forces—then what kinds of rhythms might these unknown forces produce?

Unknown they might be, but at least some guesses could be made about their nature. Leaving out of consideration for the time being such forces as might originate in outer space, there are many forces that are associated with the earth's immediate environment that could be expected to fluctuate with known cyclical events. The obvious ones were the daily, the lunar-tidal, the lunar, and the annual cycles. Strangely, the biological effects of most of these had never been carefully investigated. Frank Brown set himself the task of surveying the field, and it led him to the longest series of experiments that he has carried out with just one organism. By 1970 he had been working with it for more than ten years, and was then looking forward to further experiments projected to last for at least another five. That organism was the potato.

Why, of all things, the potato?

Brown chose it for a number of good reasons. First, he wanted to investigate the rhythm of some physiological process that was common to all living things; not like the color change of crabs or the opening of oysters, but rather some process shared by all plants and animals. Second, he wanted an organism that everyone would concede was dull; if he could show that *it* had several kinds of rhythms, then such rhythms in more active organisms would certainly seem reasonable. The potato qualified on both counts.

One common physiological process that Brown could measure in the potato (as well as in other organisms if he wished) was the metabolism rate. Recall from Sweeney's work on *Gonyaulax* that green plants both consume oxygen in respiration and produce it in photosynthesis. Brown carried out his work in the dark, so that no photosynthesis occurred, and metabolism was limited to the very slow combustion process in which oxygen is used up.

His actual experimental material was a short cylindrical plug cut from a potato, with a single eye at the top. In the dark, sickly pale sprouts grew from the eye, but the tuber contained an ample supply of stored food, and the life of the plug could be maintained for long periods in hermetically sealed containers.

While Sweeney measured the oxygen given off during photosynthesis, Brown measured the oxygen consumed by respiration, and in his case also, the amount of oxygen to be measured was minute. Tedious, time-consuming observations of barely detectable differences with the most sensitive instruments were required, and even then, Brown had to resort to statistical methods for the analysis of his data.

Still, he overcame the experimental difficulties, and he discovered some truly amazing rhythms. In addition to the expected precise 24-hour rhythm of metabolism, he found irregular variations that were associated with distortions of the barometric pressure cycles of the atmosphere. He found also that the potato plugs could predict what the barometer would read the day after tomorrow, *even though the potato plugs were maintained in hermetically sealed containers!*

To reach this conclusion he measured the day to day changes in the rate of metabolism of the potato plugs by measuring the rate at which they consumed oxygen to the millionth part of a drop; he then compared the changes in metabolism with the concurrent changes in barometric pressure.

As Brown put it: "The potato, in the form of its daily metabolic fluctuation, even in constant conditions of *pressure,* through being hermetically sealed in rigid containers, is, in effect, informing us through its metabolic changes what the weather distortions were in the regular pressure cycle of yesterday. And, whatever the means by which this information reaches the potato, the same information is providing the potato, figuratively speaking, with information as to what the weather-associated, barometric pressure will be the day after tomorrow, by the height of the afternoon peak of metabolic rate. This is tending to trace out the form of the baro-

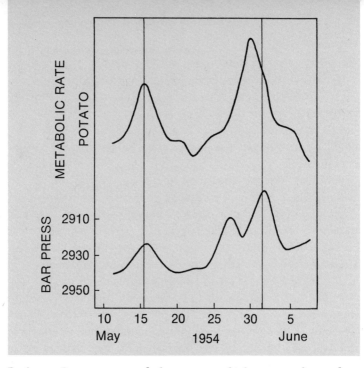

Professor Brown measured the rate at which potato plants—kept in rigid, hermetically sealed containers—use oxygen. Even though these potatoes were completely isolated from the atmosphere, their rates of metabolism corresponded to the barometric pressure in a remarkable way: they showed what the barometer was yesterday, and predicted what it would be the day after tomorrow.

metric pressure changes but doing so two days in advance. This is illustrated by the month-long study of potatoes . . . shown in the figure. Notice how [the potato] generally varies inversely with the barometric pressure change (bottom curve), but tends to anticipate the pressure changes by an average of two days. In fact, every living thing studied in our laboratory during the past 3 years—from carrots to sea-weed, and from crabs and oysters to rats—has shown this capacity to predict very safely beyond chance the barometric pressure changes usually two days in advance. It is interesting to contemplate the problem of a meteorologist sealed, incommunicado, for weeks or months in constant conditions, and asked to give

2-day weather predictions—or, for that matter, even to tell you the weather today."[7]

He also found clear relations between metabolism rate and the time of the month, and also the time of the year.

Brown measured the rate of metabolism of fiddler crabs too, and showed that their expected daily rhythm varied inversely with fluctuation of the cosmic radiation that falls on the earth from distant outer space!

In this strange field, Brown had discovered some very strange things indeed. And it was all too much for those who had been concentrating on the search for the *internal* clock. Some of them set forth to ferret out any possible slips that Brown might have made, and if possible, to expose them. They knew that the amount of oxygen that he had measured was so small that he had been forced to resort to statistical methods to demonstrate his findings. They also knew that statistics can be exceedingly tricky to handle, and that they can easily trap the unwary into false conclusions. Since Brown could hardly claim expertise in statistics, perhaps he had fallen into just such a trap.

Behind the scenes began a fantastic episode.

Not long after Brown published his first studies on potato plug rhythms, there appeared in the journal *Science* one of the most curious papers to reach print since science came of emotional age. It was entitled, "Biological Clock in the Unicorn,"[8] and it was written by LaMont C. Cole, who was a professor of zoology at Cornell. Cole seemed eminently qualified to write the paper. He held a Ph.D. degree from the University of Chicago, he had been appointed to a full professorship at Cornell in 1953, he had specialized in the mathematical analysis of populations, and he was a member of the Biometric Society and the American Statistical Association. In his paper he proceeded to show by "experiment" that he could take

---

[7] Frank A. Brown, Jr.: "The Rhythmic Nature of Animals and Plants," *American Scientist*, Vol. XLVII (June 1959), pp. 164–86.

[8] LaMont C. Cole: "Biological Clock in the Unicorn," *Science*, Vol. CXXV (May 3, 1957), pp. 874–6.

numbers from a table of completely random numbers, and by a little statistical manipulation, arrive at another series of numbers which, when plotted on a graph against "hours" showed a clear rhythmic pattern. The similarity between Cole's curve and many, many others that had been found in living clocks was remarkable.

One special effect of Cole's paper was to impugn many of Frank Brown's claims as mere artifacts—that is, as not being real, but falsely produced by his statistical methods.

Of course the satire in the title, "Biological Clock in the Unicorn," was not lost on biologists. What more amusing than to discuss with a straight face a nonexistent clock in a mythical creature?[9]

Actually, nowhere in Cole's paper was Frank Brown's name mentioned, but the innuendo was clear enough to everyone in the field, and for years Brown could not attend a scientific meeting without hearing *sotto voce:* "Unicorn!"

"That paper," says Brown, "was really below the belt. It got us deprived temporarily—for two or three months—of all support for our research. While Cole never quoted our work, everyone said, 'here is something that will stop 'em before they get off the ground.'"

Then in the summer of 1964, while Brown was working at Woods Hole, there arrived a young physiologist from the University of Strasbourg; his name was A. Heusner, and in addition to being a physiologist, he was an expert mathematician. He came to see Brown, and told him that he was interested in some work that F. H. Barnwell had done on energy metabolism in chick embryos. But he was worried about Barnwell's conclusions because they were based on statistical analyses, and all such work had been called into question by Cole's paper. He also told Brown very frankly that he suspected that Brown's own recent conclusions might represent nothing more than the "Cole artifact."

---

[9] Cole undoubtedly knew that the unicorn in myth was regarded as a symbol of virginity, but it seems doubtful that he meant to imply that his own unicorn might finally prove to be sterile.

"Well," said Brown, "it's your prerogative if you want to maintain that. We think we're right, but if you can prove otherwise, go right ahead."

Then Heusner said: "Incidentally, is there a copy of the Cole paper anywhere around?"

And Brown said: "Sure, you'll find it in the library."

Brown now recalls: "When Heusner came back about three hours later, he was just fit to be tied. 'That Cole paper has nothing to do with your work, or Barnwell's either. As a matter of fact, I wasn't worried about Cole's paper just because of your work. If anyone can take a table of random numbers, and with just a casual manipulation of those numbers —in a random method so-called—now get five chances in a hundred probability, why this is impugning the whole of what I thought was good mathematical logic. But by the time I got halfway through Cole's paper, I knew that he was going to get a cycle, and I knew what kind of cycle it would be. He had put that cycle onto random numbers himself, and without even knowing it; because he did not follow his own random equation!' "

And Heusner reported exactly that at the Feldafing Summer School.[1] It did not endear him to Brown's "opponents," but it did serve to clear the air.

"For six or eight years," Brown says, "people had been clobbering me with Cole's paper. I was not enough of a mathematician to see the flaw in it. So it was a tremendous relief when young Heusner got that thing off my back.

"And then after all this, it pleased me no end the other day to hear that I had been named a national lecturer for Sigma Xi. Because this is a little bit of approbation—this said in effect: 'You can now go out and talk without ostracism to the respectable scientists of America.' "

Frank Brown did go out and talk to them about the effects of subtle geophysical forces on biological rhythms. He told them very candidly that he had no idea what all these

---

[1] A. Heusner: "Sources of Error in Study of Diurnal Rhythms," in *Circadian Clocks* (Amsterdam: North-Holland; 1965), pp. 3–12.

forces might be, but that at least one of them could be sensed by at least one organism. He and his colleagues had been able to show that a certain marine mud-snail, *Nassarius,* can perceive changes in the strength of a magnetic field as weak as the magnetic field of the earth.[2]

Michel Gauquelin, who has written extensively on astrology, and who has studied the relations between cosmic and biological phenomena for many years, has sympathetically described Brown's work with mud-snails in these words:

"Brown and his colleagues began their studies with a small mollusk called *Nassarius,* which resembles a slug and lives in puddles on the beach. These animals were chosen because of their slowness. The experimental setting . . . was simple but original. The mollusks were placed inside a 'corral' containing two centimeters of water. They could leave the container through the neck of the corral, but only one at a time. As they left the enclosure, a fan-shaped indicator permitted the experimenter to measure the angle of the direction taken by the animal in leaving its prison. In this fashion the exits of thirty-three thousand *Nassarius* were observed; some of the slugs turned left, some right, and some continued straight ahead. When the researchers submitted the pattern of exit orientation to mathematical analysis, they found that the direction in which the *Nassarius* turns upon leaving depends on the time of day. In the morning the mollusks usually turn right; at other times they turn left more often. Certain components of terrestrial magnetism also change in the course of the day. Continuing their experiments through the summer of 1959, Brown and his associates discovered that the mollusk's rhythm of orientation is affected by the phase of the lunar month just as the sensitive registering needle of the geophysicist is."[3]

Brown made no claim that the fluctuations in the earth's

[2] Frank A. Brown, Jr., H. Marguerite Webb, and F. H. Barnwell: "A Compass Directional Phenomenon in Mud-Snails and its Relation to Magnetism," *Biological Bulletin,* Vol. CXXVII (1964), pp. 206–20.
[3] Michel Gauquelin: *The Cosmic Clocks; From Astrology to a Modern Science* (Chicago: Henry Regnery Co.; 1967), p. 139.

magnetic field were actually responsible for *timing* mud-snails, or any other organism; but he had definitely shown that they could perceive such fields. That being so, then they might well also be able to perceive other geophysical forces—however subtle, and however obscure.

How then to learn more?

Granting that subtle forces may influence the timing of some organisms, how might one visualize experiments that could reveal their nature?

A logical approach would be to seek out an environment in

When the mud-snail Nassarius *is permitted to leave an exit facing south, the position of either the sun or the moon appears to affect its direction of travel. If the strength of the magnetic field to which the snail is exposed is increased, the response of the snail is increased. The changes of direction indicated by the arrows are exaggerated.*

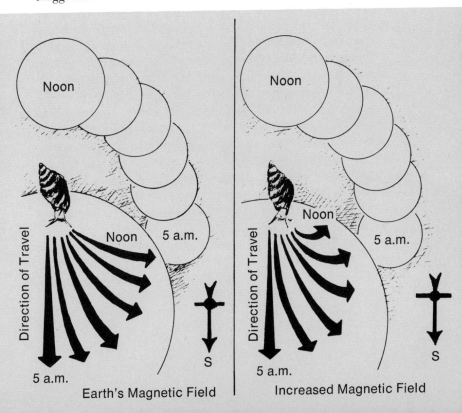

which all such forces would be absent. Just as de Mairan removed his sensitive plant from the influence of light by putting it into a dark closet, so might Brown remove his potato plugs from forces in the earth's environment by putting them into orbit around the earth; or even better, by sending then into deep space.

Excellent! In principle. But how could he get back the necessary information about the rhythm of the potato? If he based his space experiments on the potato's respiration, on which he had already accumulated such an extensive body of data, then he would need a respirometer instrumented to send radio signals back to earth. Radio transmission presented no special problem, but the respirometer itself was a very different matter.

Recall that Sweeney's Cartesian diver depended on gravity for its action. Such a respirometer would not work in a weightless environment. And there was only one other suitable kind of respirometer that had ever been devised. That depended on the forces at an interface between air and water, and Brown was sure that in weightlessness that would not work either.

A respirometer based on an entirely new principle was demanded. Of course great strides have been made by instrument engineers, particularly since the 1950's, and Brown felt sure that a suitable respirometer could be developed—given the time, and more critically, given the money. He knew that the National Aeronautics and Space Administration was vitally interested in all areas of bioscience research in outer space, and accordingly he applied for a grant to develop a new respirometer. This was approved, and as of the post-Apollo-13 period Brown was busy streamlining and refining the design, and looking forward eagerly to the day when he can plot a curve of the metabolism of a potato in outer space.

Government grants for research are well publicized, and when Chicago newsmen got wind of the fact that a biology professor at Northwestern planned to send a potato into orbit around the earth, they promptly dubbed the project "Spudnik" and gave it a great deal of feature coverage. They did not neglect the controversial nature of the idea, either.

Controversial it may be, but that has not kept Frank Brown from becoming a recognized world authority in his field. He wrote the article on biological rhythms for the 1966 edition of the *Encyclopedia Britannica;* he wrote the article on periodicity in organisms for the 1967 *McGraw-Hill Encyclopedia of Science and Technology;* and he wrote the article on endogenous rhythms for the Reinhold *Encyclopedia of Biological Sciences* edited by Peter Gray (1961). He has been an editor of *Physiological Zoology,* a trustee of the Marine Biological Laboratory at Woods Hole, a member of the corporation of the Bermuda Biological Station, and of the Mt. Desert Biological Laboratory. He holds memberships in many scientific societies.

How does Frank Brown view the status of the living clock problem today?

"To begin with," he says, "you have to concede that just too many things have been discovered in recent years to be explained by an internal rhythm alone. The endogenous rhythm by itself is not enough.

"Sure, I can take my accurate 24-hour internal rhythm, and I can convert it to a different rhythm—say a rhythm of 24 hours and 17 minutes. That is, I can do it if you give me enough incentive. Suppose you agree to pay me $50,000 two weeks from now if, every day, I appear in that doorway just 24 hours and 17 minutes later. I'll get it! I will have turned myself into a frequency transformer, and that is a simple matter—even for a cold mechanism, let alone for a living one."

He swiveled his chair around and studied a series of curves mounted on his office wall. "Look at the rhythm of the sprouting of that seed. It is an annual rhythm. Is anyone going to tell me that inside that dry seed there is a circadian clock so arranged that its alarm goes off once a year? I don't believe it. And then all these other rhythms that have been found in plants and animals. Think of them as tapes on a computer. Here's a tape with an exact 24-hour cycle. Here's another that is set at 24 hours and 50 minutes. And then a semi-lunar tape and a lunar tape and an annual tape. Suppose I switch from one tape to another at some instant in time. Or suppose I erase some of them. See how very complicated it all is? And my 'opponents' are trying to tell me that all these tapes are governed by just one circadian tape! I simply don't believe it."

Riffling through a stack of papers on his desk, he drew out a manila folder. "Now here's a manuscript I just sent off to *Oceanology International.* I called the article 'The Solunar Clocks of life.' I thought I might as well call them what they really are. They are not solar. They are not lunar. They are *both.* They are 'solunar.' In the article I said that the time has passed . . . Here it is, in the closing paragraph: 'The time may well come when we shall be able to establish beyond reasonable doubt the existence of autonomous timers approximating some or all of the calendar-clock periods. However, the time when all the observed properties of endogenous biological rhythms could still be explained exclusively in terms of such timers has definitely passed. The evidence at hand compels us to conclude that the solunar clocks of life are either dually timed—i.e., by internal timers *and* pervasive, subtle, external ones—or simply, singly by external ones alone.' "[4]

Frank Brown freely concedes that this is not a view with which most biologists agree today. "You come along with a new idea, and it meets a wall of resistance. It takes five to eight years before people begin to listen to what *I* say in any of my papers. Traditional biologists have terribly fixed ideas, and it takes an awfully long time to un-idea them."

If he had it to do over again, would he take any different course?

"Oh, no, no, no. We went into strange things for the fun of it, and some very strange things came out. Maybe it was a little like opening Pandora's box. All kinds of things came out of that. And a lot came out of our box too. But there was one thing that didn't get out, and we still have it.

"That is hope."

[4] A well-rounded summary of Brown's present position, in terms not too technical for the interested general reader, is available in his paper, "A Unified Theory for Biological Rhythms," *Circadian Clocks* (Amsterdam: North-Holland; 1965), pp. 231–61.

16. Hamner: Living Clocks at the South Pole

On a mild fall evening in 1957, Karl C. Hamner, professor of botany at the University of California, Los Angeles, propelled his then ample frame through the kitchen door of his home in West Los Angeles, two frosty bottles of Hamm's beer in one hand, and two glasses in the other. Pouring, he built two foamy collars in the glasses, lost himself for a moment in contemplation, and then turned with a smile to one of his most imaginative graduate students—James C. Finn, Jr.

Recalling that evening a dozen years later, Professor Hamner told me about the brainstorming session that he and Finn had had, and I was again impressed with the unexpected ways in which research ideas can be born. "To this day," Hamner told me, "neither of us can say who had the idea first—nor do we care."

The problem, and the idea for its solution, had to do with whether living clocks are timed internally, or whether they are timed by Frank Brown's subtle geophysical forces.

Hamner was inclined to think that Brown might very well be right. Certainly Brown's graduate students at Northwestern were driving hard with that as their working position. They had proved beyond question that they could change the direction that a mud-snail traveled by turning a magnet close to the snail. This could happen only if the snail were somehow able to sense the direction of the magnetic field.

Finn, an attractively clean-cut young man, sipped his beer and considered the problem. The mud-snail bit, he suggested, was very hard to believe. Some of Brown's conclusions had been questioned before—on statistical grounds—but even if the snails could sense the field of a strong magnet, how could they sense a magnetic field as weak as that of the earth? That field was hardly half a gauss, a small unit. Even a dime store magnet was five to ten times stronger than that. How could a primitive form like a mud-snail perceive minute changes in a force that was itself so very small?

Hamner eyed his well-dressed young graduate student with approval. He liked the way Finn argued—with incisive logic, but reasonably. His argument about the snail surely seemed reasonable. Yet Hamner knew that Brown was fully confident of his position. "And his graduate students at Northwestern are too darn smart," he added, "to accept any position that doesn't have strong experimental support."

Besides, Brown wasn't arguing that it was exclusively magnetic fields that controlled living clocks. He was only showing in the laboratory that one particular organism can be controlled by that particular force. That being so, it was reasonable to suppose that *some* external force—perhaps unknown —*did* control them.

What other forces were there besides magnetism? Well, variations of electrical charges in the air, like Southern California's Santa Ana winds. And gravity, obviously. Plus other forces that originate in outer space, such as those caused by sunspots.

And as they sipped their beer they pondered the problem of how an organism might be tested so as to be completely free from any forces that were associated with the rotation of the earth. Suddenly both had the same idea at the same time. As Hamner later wrote, in a calmer mood: "There are two places on the earth, the geographical poles, where all diurnal variables can be essentially eliminated. Organisms could be placed on turntables which, if revolved counter to the earth's rotation once each 24 hours, would completely eliminate diurnal rotation. In such circumstances it would be inconceivable that organisms could receive any information of a diurnal nature from the external environment, except that which might be accidentally introduced by the experimenters, and that which might be associated with the daily 'rotation' of the magnetic pole about the geographical pole."

But now the two brainstormers were in no calm mood. They leaped to their feet and raised their glasses high. Then they sat down. For they quickly realized the difficulties of getting a quantitative biological laboratory to either pole. How could plants and animals even be kept alive? Would it be practical at all? And even if it were, it would require a prohibitive sum of money.

But no money was in sight, and the idea languished. A year passed. Two. Nearly three. James Finn earned his doctorate and joined the Space and Information Systems Division of North American Aviation. Karl Hamner raised cocklebur plants and Biloxi soybeans for generation after generation in his greenhouse on the UCLA campus, and continued his search for the elusive hormone that he knew caused his plants to flower.

Then one day, as he was leafing through the journal *Science,* Hamner's eye lit on an inconspicuous announcement. It said that the U.S. Antarctic Research Program of the National Science Foundation was soliciting independent proposals for research projects that could only be carried out in a polar environment.

At once the experiments that he and Finn had thought of that fall evening nearly three years earlier flashed through his mind. South Pole, North Pole, what difference would it make?

The results should be the same. If the National Science Foundation wanted research proposals, they were going to get one from the Botany Department at UCLA.

So in January, 1960, a proposal for a research project on the rhythms of plants and animals at the South Pole appeared on the desk of the Special Assistant for Antarctic Affairs in Washington.

However, well before Karl Hamner could expect any word from Washington (for it takes many months for agencies there to process research proposals), he received a letter that surprised him, from a far different source. The organizing committee for the Twenty-fifth Symposium on Quantitative Biology of the Long Island Biological Association was asking if he would attend and present a paper. They thought that the international audience there would be much interested in hearing about his work on how different light-dark cycles affect the daily rhythms of plants.

Karl Hamner reread the letter. The symposium was entirely on biological clocks, and their behavior was interesting him more and more. He would go. All of the world authorities on rhythms would be there, and it would be a good chance to find out what they would think of the South Pole project. He decided that if they approved of it, he would go, if the grant were forthcoming. If they did not, he would withdraw the proposal.

On Long Island, on that June day in 1960, Karl Hamner listened intently as Frank A. Brown, Jr., Morrison Professor of Biology at Northwestern, drew to the conclusion of his complex and controversial paper, "Response to Pervasive Geophysical Factors and the Biological Clock Problem." In summary, Brown said: "There are good reasons to believe that the living organism relies heavily upon a geophysically-dependent clock-system. Indeed, since the organism is such a delicately poised physico-chemical entity as to reflect small fluctuations in both identified, and probably still unidentified, subtle geophysical factors, the question arises as to how essential to life itself are the various components of the natural geophysical complex of the earth's atmosphere. If living things have evolved to depend in their orientation in time and space upon some of them, to what extent and in what ways can the environment become different without effecting radical changes in the total picture of life on our planet?"[1]

As Brown finished and sat down, a man of medium height and weight, with light hair combed straight back and dark-framed glasses resting on his prominent nose, rose to comment on the paper. It was Professor Erwin Bünning, of the University of Tübingen in West Germany. For many years he himself had fought the active opposition of many scientists to a hypothesis of his own. But in spite of that he had no choice but to point out facts from his own experience that did not agree with Brown's conclusions.

Speaking in precise English, accented by his native German, he cited specific experiments with plants and said that nowhere in these experiments had anything been found that would suggest that the plants were being affected by any external factor. In these cases, and in others that he could cite, he could see no reason even to look for such factors. Still he felt obligated to add that the absence of external factors observed in many cases did not prove their absence in *all* cases.[2]

When Bünning sat down, Karl Hamner rose to present his own suggestion:

> I believe I can present an experimental approach which will provide a definite answer to the problem under discussion. About three years ago, it occurred to Dr. J. Finn, one of my students, and myself that the rotation of the earth per se may be considered as an environmental variable. If an organism

[1] Frank A. Brown, Jr.: "Response to Pervasive Geophysical Factors and the Biological Clock Problem," in *Cold Spring Harbor Symposium on Quantitative Biology,* Vol. XXV (1960), p. 69.
[2] Even the most violent opponents of Frank Brown's ideas must qualify their arguments to this extent. In pure logic, Brown's position is unassailable. A universal negative is impossible to prove even of a force that is clearly identifiable, to say nothing of a force that is not.

in some way could "understand," or in some way measure, the earth's rotation, the organism would have a very accurate clock or timing device. For example, let us say that organisms are receiving, constantly, some kind of radiation from a fixed point in space. Then, as the earth rotates, the direction of this radiation would change, and by measuring the direction of the radiation in relation to gravity, the organism would be able to tell time accurately. I am using this purely as an example and not postulating some unknown radiation from a given direction. However, operating upon such a postulate, we set up some very preliminary experiments in which we placed plants on a horizontal position and rotated them so that the direction of the gravitational force was constantly changing with respect to the plant, and, therefore the force of gravity as a factor was essentially removed. Our preliminary results indicate that such plants may lose their sense of timing since the rhythmic leaf movements seem to disappear immediately. The results are very preliminary and of course inconclusive.

In January, 1960, we proposed to the National Science Foundation a project to be performed at the South Polar region of the earth. It was proposed that organisms which exhibit endogenous rhythms be placed on turntables near the Pole. If such a table were adjusted to rotate once each 24 hours in a direction counter to the rotation of the earth, then the organism would remain still with respect to a fixed point in space. Furthermore, at the polar region one would not expect diurnal fluctuations of the environment. In other words, such an experiment would remove the rotation of the earth as a factor in the timing mechanism. If such an organism lost its time sense, this would prove that in some way the organism is "aware" of the earth's rotation. On the other hand, if the rhythms are not affected, the experiment should prove that the organism is not "aware" of the earth's rotation, and that the clock is probably controlled entirely by some endogenous mechanism. Turntables could be adjusted, of course, to rotate either with the earth or against the earth at different rates. It seems to me the results would be of basic significance to our understanding of "biological clocks."[3]

[3] Brown: "Response to Geophysical Factors," p. 70. Used by permission of the Long Island Biological Association.

In response, Frank Brown said that positive results of such an experiment would be most exciting, but that if negative results should be found, they should be interpreted with caution in view of the complex daily movements of the magnetic pole around the geographical one.

Of course Karl Hamner was well aware of this complication. The magnetic pole is hard to define in the first place, and to say just where it is at any moment is even harder. Still he was convinced that any effects that might arise from this variable would be trifling, and he already had in the back of his mind an experiment that would answer that question for whatever it might be worth.

Between formal sessions, he chatted with many authorities on rhythms and received enthusiastic encouragement to go ahead with his South Pole program. "Next to taking a laboratory into outer space, your answers should be the best we could hope for. You can go after them now, and a manned space ship may never penetrate the Van Allen radiation belt."

When he returned to Los Angeles that June, Karl Hamner was acutely aware of the fact that if his research proposal were approved and funded, he would have only a few short months to make all preparations to place a fully equipped biological laboratory at the South Pole by December at the latest—midsummer in the Antarctic.

To this task Karl C. Hamner brought a wealth of know-how and experience, in the administration of research as well as in academic affairs. He had studied at the University of California at Los Angeles, at Fresno State College, and at the University of California at Berkeley, where he received his B.S. degree. He then entered the University of California Medical School and held out until he had finished the first year, even though he had become increasingly aware that he was deriving very little pleasure from it. The time had come to ask himself if he really wanted to be a doctor after all. The answer was clearly *no*. Looking back, he told me with a twinkle: "Now don't jump to the conclusion that I flunked out of medical school. Matter of fact, I was second in my class. I just realized that I didn't like all that blood."

The next fall found Karl Hamner enrolled as a graduate student in botany at the University of Chicago. He wanted to work with living things, he was deeply interested in research and teaching, and a career devoted to studying the physiology of plants held the greatest appeal for him. Upon earning his M.S. in 1934 he was awarded a fellowship, and he received his Ph.D. in botany in 1935. The following two years he spent as a plant physiologist with the U.S. Department of Agriculture in Beltsville, Maryland, and then he returned to the University of Chicago as instructor of botany. Later, he accepted a post in the government's Plant, Soil, and Nutrition Laboratory at Ithaca, New York. In 1946 he was appointed director of that laboratory.

In government facilities, as well as in industry, it is a common observation, however, that while scientists often make first rate administrators, a good many of them do not like the jobs very much, and would much prefer to be back in the laboratory. It was Karl Hamner's good fortune that he was offered a choice. The University of California at Los Angeles sought him out as professor of botany and plant physiologist in the Experiment Station of the College of Agriculture. Here was his chance to leave the desk of budgets, and personnel problems, and endless conferences with government functionaries, and return to the laboratory and the classroom. "Salary was no problem," he told me. "So what could I lose?" For more than twenty years Karl Hamner has lived and worked happily on the UCLA campus at Westwood.

But there was no air of academic calm about him on that July day in 1960 when he learned that the South Pole project was approved. Calling the staff people whom he had chosen as best suited for the venture—Sirohi, Hoshizaki, and Carpenter—he reviewed the task that faced them.

The National Science Foundation had provided a grant of $25,000 for the expedition. Within that sum, and in a very short time indeed, they must prepare to take a complete quantitative biological laboratory to the South Pole.

The basic preparations involved assembling specimens of the five different living materials that they would place on their turntables at the Pole, and in making ready the equipment and instrumentation that they would need to keep their organisms in good health, and to make records of their rhythms of activity. Each of the organisms that they had chosen possessed a rhythm pattern that was well known from extensive studies in other laboratories, and the experimental procedures for recording these were well established.

Syrian golden hamsters would be placed in cages provided with exercise wheels connected to a recorder, so that periods of activity would appear as black bars on a moving chart, and periods of rest would be a blank. Bean plants would be grown from seeds and the leaf movements of the seedlings would be recorded on infrared film loaded in a time-lapse camera. The fungus *Neurospora crassa* would be grown in glass tubes filled with culture medium, and the daily movements of the colored bands would leave their own record of activity; this would be recorded by direct observation, and also photographed. Fruit flies would be placed in bottles where they would lay their eggs in preparation for the instant of emergence of the next generation of adult flies. These would be counted by shaking them into recovery flasks and counting them after deadening them with ether. Cockroaches would be provided with delicate running wheels connected to recorders; if the wheels did not work well, the running activity of the roaches could still be recorded on the time-lapse camera.

Even the turntables had to be designed and built, together with the mechanisms for driving them. For constant speeds of rotation, synchronous motors would be used, and in addition to the primary condition of rotation of the earth's speed in the counter direction, other speeds and directions were planned. Sets of gears would be needed so that the turntables could be spun either forward or backward, and at speeds providing one rotation in 12 hours, and in one, two, four, six, or eight days.

Carpenter, who was an excellent organizer and planner, although he did not go to the pole, handled many of the preparations for the venture. He was especially concerned about the time-lapse camera. What if the intervalometer on the camera went haywire? What shop equipment would be available at the pole, for making repairs?

None at all, Hammer told him. The team would have to be completely self-sufficient. The National Science Foundation would arrange for their transportation and their food, but nothing more. Whatever else they might conceivably need, they would have to assemble and take with them. Even the commonest tools. Every hammer. Every screw and every screwdriver. Plastic greenhouse, and collapsible aluminum poles to hold it up. A bolt of photographic dark cloth, to provide periods of constant darkness inside the greenhouse. Artificial soil, and artificial plant nutrients, because the soil at the South Pole is buried under snow and ice. Film, photographic chemicals, thermometers, and every kind of recording apparatus.

Still, in a very few months they were ready to go.

As the plane of the U.S. Military Air Transport Service prepared to land in Honolulu, it carried a strange cargo. Karl Hamner's team watched over the specially designed temperature-controlled shipping containers that held their live experimental materials. These, and all supporting equipment from nails to intervalometer, came to nearly two tons. At the touchdown, the scientific team began preparations to take the precious living materials to the hotel for the overnight stay.

But the shipping containers and their living contents were never to leave the plane. An inspector of the Hawaiian Department of Agriculture had boarded the plane, and he refused to permit the cargo to be moved. Understandably, the islands must be protected against the entry of new plant and animal diseases, and the islanders maintain a very tight control. It did not matter to the inspector that Hamner assured him that all his materials were disease-free, nor that the National Science Foundation and the State Department had authorized the flight plan. The hamsters, and the cockroaches, and the fungi, and all the rest of it from the mainland would not touch Hawaiian soil.

Hamner was frantic. The temperature in the plane, now standing in the hot sun, was rising fast. Soon it would cross the hundred-degree line, and in that heat, the specimens could not survive for long.

Just then a young naval officer came up, and Hamner told him his troubles. The inspector was just doing his job, the officer pointed out. But taking the gear off the plane was not the only answer. He would rig an air conditioner on the plane, big enough to cool it down to seventy degrees in ten minutes. So ended one of the worst moments that Hamner had on the entire trip.

The next stop was at Christchurch International Airport, some forty-five degrees below the equator on New Zealand's South Island. There too, the plane was met by an inspector, this time representing New Zealand Health, Immigration, and Customs. But he was more lenient than the Honolulu man, and he told Hamner to take his shipping containers to the hotel, where he could maintain a check of their condition. But be doubly sure, he warned, that none of the animals got loose.

Secure in their hotel, Hamner, Sirohi, and Hoshizaki surveyed the containers with satisfaction. The thermometers stood at a steady seventy degrees.

They would stay in Christchurch just long enough to reset the hamsters and the fruit flies to New Zealand time. They would turn on the lights for 12 hours beginning at 8:00 A.M., and turn them off just 12 hours later. After a few days of this entrainment, the animals would be prepared to maintain that rhythm in the continuous darkness in which the experimenters would keep them at the Pole. For the *Neurospora* fungus, they would make up fresh cultures in Dr. Wohlschlag's laboratory at McMurdo.

All seemed well in hand until someone glanced at one of the cages. One of the hamsters had escaped. A mad search followed, into corners, under beds and into closets, until someone found the escaper, and thrust him back into his cage.

Karl Hamner sank to the edge of a bed with a sigh of relief. He was ready to settle for a meal and a bottle of Commonwealth ale.

The next stop was at McMurdo Station on the low volcanic hills of Cape Armitage, Ross Island. McMurdo serves as the "primary logistic facility" from which the U.S. Antarctic Research Program conducts plane flights to supply the inland stations and to support science programs in the field. Also, following the International Geophysical Year, a year-round science program was set up at McMurdo. There work is in progress on biological research, monitoring of cosmic rays, and observations of satellites. From McMurdo, Hamner and his men pushed on for the South Pole itself—all but a disappointed Hoshizaki, who was assigned to stay at McMurdo to make experiments paralleling those to be done at the actual pole. He relieved the blow by growing a beard, and "coming out," Hamner says, "looking like a villain in a war movie."

As the Navy plane crunched down on the snow at South Pole Station, Hamner jumped up, ready to wrestle plant and animal containers across the snow and into the heated garage building that awaited them. As he dragged a box clear of the plane, the resident Navy doctor walked up and welcomed him. But he warned him against overexertion and told him to take it easy with the box. The elevation was ten thousand feet right there, and the air was thin. And, he said, the temperature was forty below zero.

Hamner was jarred. Forty below, even in polar summer. He would have to rush the containers under cover before the sub-zero cold could penetrate to their living cargo. But with all hands turning to, every item was soon safely stowed in the old garage building. The air, tempered by diesel heaters and circulated by electric fans, held an even seventy degrees to within two degrees up or down.

With all secure, Hamner and his team began unpacking their two tons of gear and planning the layout of their experiments. Immediately they realized that the garage floor was cold, and they placed all of the live materials on table tops three feet above the floor.

They assembled the collapsible aluminum poles and spread plastic sheets above them so that they could maintain controlled temperature and humidity within. They cut lengths of photographic dark cloth to provide for continuous darkness when they needed it. They placed high-intensity fluorescent lights overhead. They planted bean seeds in moist vermiculite and set them aside to sprout. They wrapped glass tubes of the colorful *Neurospora* fungus in dark cloth so that the progress of the colored bands along the tubes would continue in total darkness. They assembled the delicate exercise wheels for the cockroaches, and checked over the recorders for the hamster treadmills. They checked the fruit fly culture in their half-pint milk bottles, and made a random selection of those that would be used for experiments. The rest they reserved for controls. They set up the turntables, and checked out the synchronous motors and the gears. Within a week after they had landed at Pole Station, they were ready to start their first experiments.

With all lights out inside the plastic greenhouse, they opened the time-lapse camera and loaded it with infrared film. They set the camera's intervalometer to take one picture every twenty seconds of the leaf positions of the bean seedlings. Then they set the turntables in motion and turned on the infrared light source.

Anxiously they awaited the development of the first strip of film. *It came out completely black.* That was strange. They had used the same shutter speed and diaphragm opening that they had carefully checked in Los Angeles. But the film was grossly overexposed.

And every day that followed throughout that frustrating week yielded nothing but jet black negatives. Baffled, they closed the lens diaphragm to its smallest opening, and set the shutter speed at its fastest. When they developed that film they saw again a completely overexposed row of frames. Somehow, infrared was getting to the film, and it was not coming through the camera lens.

So they sat back and reviewed the sequence of operations step by step. The standard procedure was to take the unloaded

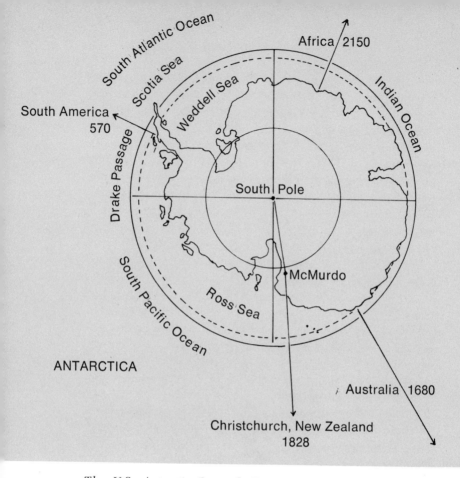

South Atlantic Ocean

Scotia Sea

Weddell Sea

Africa /2150

Indian Ocean

South America
570

Drake Passage

South Pole

McMurdo

South Pacific Ocean

Ross Sea

ANTARCTICA

Australia \1680

Christchurch, New Zealand
1828

The U.S. Antarctic Research Program of the National Science Foundation provided the facilities for critical experiments on biological clocks at the South Pole. The organisms and equipment for a complete biological laboratory weighing two tons were flown from New Zealand to the USARP Station at McMurdo, and from there to the pole.

camera into the greenhouse and mount it in place, to open it, unwrap fresh film, load the camera, and then close it tight. After making test exposures they would open the camera, take the film out, wrap it up, take it into the darkroom, and develop it at a rigidly standard time and temperature.

And time after time the film was hopelessly overexposed. At some point in the operation infrared was getting to the film. Was it possible that in the dark greenhouse, with no infrared light turned on, there was still some level of infrared radiation there? A simple test would give the answer.

In the photographic darkroom itself, known to be free of infrared, they took a strip of fresh film and wrapped it in opaque foil. Then they took the package into the darkened greenhouse, unwrapped the film, and let it stand for a few minutes. Then they wrapped the film up in the foil again, took it back into the darkroom, and developed it. It came out beautifully clear. No ray of infrared had reached it.

The answer was now obvious. The photographic dark cloth, which served well enough to keep the experimental animals and plants dark, was leaking infrared like a sieve. And the simple cure for that was to load fresh film into the camera only in the regular photographic darkroom.

From that day on, the camera made perfect records. In the experiments with beans, the intervalometer opened the camera shutter every twenty seconds for several days in succession, and the recording was continued until a reduction of plant vigor developed. Everyone was gratified by the sharpness of the negatives, and the precision with which the intervalometer made exposures every twenty seconds. Actually it was like time-lapse movie photography, in infrared, and when they got back to Los Angeles, they would run the film through a projector for amusement. The bean leaves, they knew, would wave up and down like birds flapping their wings.

The same photographic system was put to use to record the running activity of the cockroaches. It turned out that the exercise wheels that had been so carefully built were just too delicate to work reliably, and the camera was used instead. The film record allowed at least rough interpretation. When the cockroaches were at rest, their images on the frames were sharp, and when they were active, the images were blurred.

Within a month, Hamner and his team finished the entire experimental program that they had laid out. Four out of the five experimental materials had given clear-cut answers. The cockroaches did not, and Hamner thinks that this was due

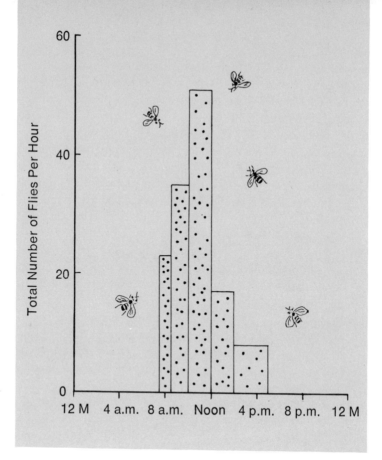

When fruit flies at the South Pole were spun opposite to the earth's rotation, they continued the normal well-defined rhythm of emergence from their pupal cases that they display in northern latitudes. This demonstrated that the rhythm is not timed by any factor associated with the earth's rotation.

partly to the team's lack of experience in working with these insects, but mostly to the fact that the particular strain they used never developed a really clear-cut rhythm at any time.

The hamsters, and fruit flies, and bean plants, and fungi all showed rhythm patterns that led to the same conclusion. It did not matter whether they were held stationary at the pole, or were spun forward, or backward, on the turntables. It did not matter whether the speed of rotation was once in 12 hours, or once in a day, two days, four days, or six days. Under all of these conditions the rhythms continued as though nothing unusual had happened to them. They might as well have been in the plant physiology greenhouse in Los Angeles.

Karl Hamner summarized the results in these words:

> Prior to the experimental program at Antarctica, one of three results was anticipated: (1) all external expressions of rhythmicity would cease because the experiments were carried out at the Pole; (2) the rhythms would be altered to new frequencies on the turntables; (3) the diurnal rhythms of none of the organisms would be influenced by either location at the Pole or by rotation on turntables at the Pole. Our data conclusively support the last possibility.
>
> From these results it can be stated that external environmental variables of a diurnal nature associated with the rotation of the earth have no detectable influence on the basic mechanism . . . of the biological clock. We conclude, therefore, that although it is possible that some external periodic stimulus may regulate the biological clock, it does not arise from any factor associated with the earth's rotation. We feel that these results give a negative answer to one particular hypothesis concerning regulation of the biological clock by geophysical stimuli. We appreciate the fact that these experiments have not added any information concerning the real nature of the clock other than to make highly improbable this one particular hypothesis concerning the regulation of the clock's mechanism.[4]

A careful examination of the records given in this *Nature* paper shows that Hamner and his team were hard at work on Christmas Day of 1960. When I asked him whether they had any relaxation at all that day, Hamner chuckled.

"Well, not on Christmas. But on Christmas Eve, everyone got a little tipsy. Some more than others. The beer was all frozen, so we had to melt it on the stove. One of the fel-

[4] Karl C. Hamner, J. C. Finn, Jr., G. S. Sirohi, T. Hoshizaki, and B. H. Carpenter: "The Biological Clock at the South Pole," *Nature,* Vol. CXCV (August 1962), pp. 476–80.

Karl C. Hamner in his greenhouse at UCLA uses cocklebur plants, foreground, *and Biloxi soybeans,* behind, *to study the hormones that cause flowering.*

lows in the dormitory forgot four cans on the stove, and while we were asleep the cans exploded and threw beer all over the room."

After the team had returned to California, one loose end remained to be tied up. Karl Hamner remembered well a Frank Brown comment at Cold Spring Harbor that negative results in the experiments would be open to question because of the rotation of the magnetic pole around the geographical pole. To answer that question, Hamner devised a particularly ingenious experiment, and carried it out soon after he returned to the UCLA campus.

Starting from the known fact that the earth's magnetic field is about a half gauss, he reasoned that changes in the location of the magnetic pole in relation to South Pole Station could have influenced the organisms on his turntables by no more than a fraction of that. If, then, he could show that much larger variations in magnetic field did not influence the rhythms of his organisms, Brown's argument on that point would collapse.

Accordingly he set up a large turntable geared to rotate at the same speeds as his South Pole turntables, and placed a powerful permanent magnet at one side. An organism on the edge of the turntable would be subjected to a magnetic field of twenty-five gauss (fifty times the earth's magnetic field strength) when it was nearest the magnet, and only two

Bean plants were rotated on a turntable past a powerful permanent magnet. When a plant was nearest the magnet, it was in a magnetic field fifty times that of the earth's field; when it was farthest from the magnet, it was in a field about the same as the earth's. Even this exaggerated variation had no influence on the daily rhythm of leaf movement.

thirds of a gauss when it was farthest from it. The same strains of bean seedlings, fungi, and fruit flies that were used at the pole were placed on the edge of the turntable. At no time did these exaggerated variations in magnetic flux disturb the daily rhythms of any of the organisms.

In the judgment of the majority of biologists, Karl Hamner had provided a conclusive answer to the subtle geophysical force question. And he had done it in six months and within a budget of $25,000!

Exciting and dramatic as it is, the South Pole story is a very small slice indeed out of the life of a man whose entire scientific career has been unbelievably productive. His colleagues hold him in the highest esteem, but personally he is modest and unassuming, and he was much surprised when he learned that Franklin D. Murphy, then chancellor of the University of California at Los Angeles, had received the following letter:

11 January 1961

Chancellor
University of California at Los Angeles
Los Angeles
California

Dear Sir:

We recently had the pleasure of working with Dr. Karl C. Hamner and his team (G. S. Sirohi and T. Hoshizaki) in his studies here at the South Pole.

The entire station personnel joins me in my thanks to Dr. Hamner. While most summer scientific personnel pose problems to the function of the station (in greater or lesser degree) Dr. Hamner was a definite asset; he more than cooperated. During the past month, we have had to shovel more than one hundred tons of snow, and Dr. Hamner joined us nearly every time. In his spare time he built bedside shelves for some of our personnel. He cheerfully joined others in dishwashing as well as taking his own turn at this duty. The station personnel showed a remarkable interest in his experiments and he was always most eager to take the time to explain them.

His pleasant personality, generosity, and willingness to turn to made him a most welcome guest, and it was a great pleasure to associate with him.

> Sincerely,
> *Philip K. Swartz Jr., M.D.*
> LT MC USN
> *Officer-in-Charge*
> *South Pole Station.*[5]

In the years since his South Pole experiments, Karl Hamner has turned a great deal of his time and attention to the problem of specifically identifying the hormone that is associated with the flowering of plants. But he still maintains a continuing interest in living clocks.

He wrote the definitive chapter about them in the most comprehensive treatment of man's views of time yet published.[6] He was chairman of the session on "Photoperiodic Time Measurement" at the Biochronometry Symposium sponsored by the National Academy of Sciences–National Research Council and the National Aeronautics and Space Administration in September, 1969, at Friday Harbor, Washington.

After listening to dozens of learned papers read by world authorities at that symposium, Karl Hamner gazed out over the sparkling blue waters of San Juan Channel toward Lopez Island and quietly told me:

"You know, our understanding of rhythms really stands about where gravitation stood before Isaac Newton. What we need now is another Newton."

---

[5] Used by special permission of the U.S. Antarctic Research Program, National Science Foundation, and Dr. Swartz.
[6] Karl C. Hamner: "Experimental Evidence for the Biological Clock," in *The Voices of Time*, J. T. Frazer, ed. (New York: George Braziller, Inc.; 1966), p. 281.

17. Pittendrigh: Review and Perspective

We have now looked at living clocks from many different points of view. Historically, from the first observation of a living clock in 1729 down to the present day; genetically, from a simple one-celled organism that can tell time—to man himself; and geographically, from Norway to the South Pole and from Japan to Germany. We have been confronted with a welter of overlapping and conflicting theories and ideas, and we have met personalities ranging from the unassuming to the tempestuous.

We will do well to pause, now, to place the highlights of these adventures of the laboratory and of the mind in summary review, to bring out their most meaningful relationships, and for perspective on new knowledge that awaits discovery.

For this insight we are privileged to turn to an outstanding authority in the field, Colin S. Pittendrigh, who is now professor of biological sciences at Stanford University.

Pittendrigh was born in Whitley Bay, England, earned a B.Sc. with first class honors in biology from the University of Durham in 1940, and, during World War II, was assigned by the British Colonial Office to attack the problem of malaria in Trinidad, B.W.I. From 1943 to the end of the war he cooperated with the U.S. Army Medical Corps, combatting malaria on Trinidad army bases. Turning, then, to graduate study in the United States, he earned his Ph.D. at Columbia University in 1948, and became a United States citizen in 1950.

In 1947 he joined the faculty at Princeton University and soon became fascinated with the problem of biological clocks. The source of his interest, as we have so often seen, came from research in another field. Pittendrigh was studying the ecological basis of malaria in Santa Catarina, Brazil, and became more and more impressed by the cyclical behavior of the mosquito-malaria relationship within the total ecological environment.

The mosquito itself does not lend itself very well to laboratory research, and Pittendrigh selected the fruit fly, *Drosophila*, as a more promising experimental insect. For the "hands of the clock" he chose for most of his studies the time of emergence of the adult fly from its pupal case. When he published the results of an intensive program of research and interpretation in 1954, biologists reacted with an upsurge of excitement.[1] At the Cold Spring Harbor Symposium on Biological Clocks in 1960, Beatrice M. Sweeney and J. Woodland Hastings put it this way:

"The realization of the unique implication and importance of the temperature-independent period of biological rhythms

---

[1] Colin S. Pittendrigh: "On Temperature Independence in the Clock System Controlling Emergence Time in *Drosophila*," *Proceedings of the National Academy of Sciences, Washington, D.C.*, Vol. XL (1954), pp. 1018–29.

emerged rapidly in the early 1950's. . . . But it was certainly Pittendrigh's elegant paper on *Drosophila* which first convincingly formulated the clock generalization with regard to persistent daily rhythms; and it was in this paper that the problem of temperature-independence was first fully and forcefully placed in what we now consider to be its proper perspective. Subsequently there rapidly evolved an appreciation for the view that rhythms are a manifestation of time-measuring, for which temperature-independence is clearly of great functional importance.

"We do not suggest that these questions had not been previously considered and appreciated, for they surely had been. But the orientation and emphasis had been different."[2]

Following such recognition, and that of a formidable volume of work that followed, Pittendrigh was appointed a full professor of biology at Princeton in 1957, and in 1965 he became dean of the graduate school there. In 1969 he accepted an appointment as professor of biological sciences at Stanford.

He has been the recipient of many honors, only a few of which need be listed here to indicate the regard in which he is held by his colleagues. He has held a Guggenheim Fellowship, and he is a member of the National Academy of Sciences and of the American Academy of Arts and Sciences. He has held offices in national scientific societies, has served on the editorial boards of scholarly journals, and has been honored by requests to deliver special lectures. He has also served on many national committees. He was chairman of the National Academy of Sciences study, "Biology and the Exploration of Mars," and from 1966 on has served as a member of the Science Advisory Committee to the Administrator of the National Aeronautics and Space Administration.

In 1965 he was invited to participate in a most unusual seminar. This was the Tenth Anniversary Air Force Office of

[2] Beatrice M. Sweeney and J. Woodland Hastings: "Effects of Temperature upon Diurnal Rhythms," in *Cold Spring Harbor Symposia on Quantitative Biology*, Vol. XXV (1960), pp. 87–104.

Scientific Research Seminar, which was held in Cloudcroft, New Mexico, and which differed from the previous specialized seminars of the AFOSR in that it spanned a broad spectrum of the sciences, and ranged from the history of science through population trends, theories of memory, combustion instability, and the frontiers of psychology, to space chemistry. "Each of the sixteen distinguished investigators who spoke at this seminar was chosen for his place on the knife-edge of his particular area of research, where work was being done that would, in all likelihood, influence the course of science."[3]

The speakers dealt with their subjects most informally in the unhurried atmosphere at Cloudcroft, allowing ample time for discussion to reach to the heart of the matter at hand. Pittendrigh's presentation on biological clocks, and the discussions that followed, occupied nearly four hours.

The Air Force Office of Scientific Research considered this first interdisciplinary approach to have been so stimulating that the contributors were asked to prepare written summaries for publication, so that a record could be made widely available to potential users.

Colin Pittendrigh's contribution was entitled "Biological Clocks: The Functions, Ancient and Modern, of Circadian Oscillations."[4] The Office of Publications of the University of New Mexico and the author have given their permission for us to refer to this brilliant discussion here. Its published form, however, is full of highly technical terms—efficient for communication among specialists, but hard going for most readers. I have therefore paraphrased the original with the aid of comments that Professor Pittendrigh gave me personally. In the interest of continuity of the text, I have made no distinction between the many passages that are directly Pittendrigh's and the paraphrase; readers who wish to check the original will find much more of interest in *Science in the Sixties*.

[3] David L. Arm, ed.: *Science in the Sixties* (Albuquerque: University of New Mexico Office of Publications; 1965), p. vii.
[4] *Ibid.*, pp. 96–111.

BIOLOGICAL CLOCKS

THE FUNCTIONS, ANCIENT AND MODERN,
OF CIRCADIAN OSCILLATIONS

De Mairan's Phenomenon

Erwin Bünning has drawn attention to a remarkable brief note by the French scientist de Mairan, written in 1729. De Mairan was evidently intrigued by the daily movements, up and down, of the leaves of certain plants. Hoping to explain the environmental causes he took a sensitive plant into a dark place where it was free of any daily cycle of light or temperature. To his surprise de Mairan found the daily periodicity of movement persisted in this essentially aperiodic environment. Recognizing the importance of his discovery, de Mairan commended the problem to his botanical colleagues. In his closing comments he anticipated slow progress in the matter —not, to be sure, because he held a low opinion of botanists but, he says, because progress in science is wholly dependent on experiment and he presumably foresaw no obvious experimental attack on his surprising discovery.

De Mairan's phenomenon attracted a long and distinguished line of botanists in the nineteenth century including de Candolle, Hofmeister, Sachs, Darwin and, especially, Wilhelm Pfeffer. Pfeffer is, of course, far better known for his discovery of osmosis and many other contributions to plant physiology than he is for two books, one in 1875 and the other in 1915, on the persistent daily rhythmicity of the movements of leaves.

To Sachs in the late 1800's it was already clear that the light cycle of the environment was not forcing any periodicity on the plant; it was only serving to control the timing of a periodicity arising from other causes. All of the botanists at the end of the nineteenth century were concerned to some extent with what adaptive functions these oscillations served. Darwin, not surprisingly, was nevertheless confident that some purpose had to exist, the implication being, of course, that to evolve as a result of natural selection some adaptive advantage is necessary.

The Endogenous (vs. Exogenous) Origins of the Oscillations

A major theme running through the literature at the turn of the century—and continuing up to 1930—was the possibility that the rhythmicity persisting in constant darkness and constant temperature arose from forces external to the plant; that some unknown factor X, an unidentified periodicity in the physical environment, was forcing the rhythmicity. Even today there is still, in fact, one laboratory that remains convinced of the reality of factor X. Professor Frank A. Brown, Jr., of Northwestern University has published a long series of studies in which he claims to detect precise 24-hour periodicities in various organic activities (respiration, movement, etc.) whose phases are closely coordinated with local time. They are then said to be "phase-fixed" to local time. These periodicities (whose reality has been questioned on statistical grounds) Brown attributes to control by factor X. His position has, however, met with considerable criticism in the field. The great majority of workers has long since concluded that de Mairan's phenomenon arises from wholly endogenous (internal), not exogenous (external) causes.

The evidence for this view derives from various kinds of observations. The principal ones are as follows: The periodicity persisting in constant dark and constant temperature can be stopped by cutting off the supply of oxygen available to an organism so that its tissues reach a state of anoxia; it can also be stopped by chilling the organism to the point where metabolism is arrested. But when oxygen is returned to the system, or the temperature is raised to a level where normal metabolism can proceed, the periodicity is resumed. The energy on which the oscillation depends is of metabolic origin; and when the oscillation is resumed, it does so at essentially the point where it stopped. Implicit in this statement is the important fact, which can be shown with many other techniques, that the phase of the rhythm in constant darkness is wholly independent of local time—that is, independent of the earth's rotation. It must, therefore, be independent of all factors that depend on the earth's rotation (including factor X).

The most impressive fact in this context is that in organisms in which it can be measured precisely, the period of the rhythm that persists in darkness is not precisely that of the earth's rotation (figure, opposite). Proponents of the endogenous nature of these rhythms argue that only by bringing forward the most cumbersome and unlikely assumptions can one explain the origin of, say, a 23 hour and 15 minute period as the product of an unknown driving cycle with a period of precisely 24 hours.

Franz Halberg has introduced the term "circadian" (from the Latin "circa," about, and "*diem*," day) for de Mairan's oscillations. This term emphasizes the important discrepancy between the oscillation period and that of the earth's rotation. It also clears out of the way a conflict of meanings inherent in the term "diurnal." Sometimes "diurnal" is used to mean "of, or belonging to the daytime," (as opposed to "nocturnal"), and at other times it is used to mean "daily." Thus the often-used phrase "daily rhythm" lacks the precision and implications of Halberg's "circadian rhythm."

Individuals within a species differ, genetically, in their free-running periods—that is, in the periods they manifest when uncontrolled by an environmental light cycle. And their periods are also open to some experimental manipulation, principally as a result of prior treatment with light.

As a matter of fact, it is now unfair to associate Professor Brown with earlier students of factor X; he no longer regards factor X as the total explanation of de Mairan's phenomenon. He recognizes that the data demand the existence of a periodicity of internal origin; his interest in factor X focuses on its potential role as a kind of pacemaker that confers precision and compensation for temperature changes on the internal oscillation.

The Ubiquity and Pervasiveness of Circadian Oscillations

De Mairan's phenomenon has been, by now, observed in a remarkably broad array of organisms: single cells, green

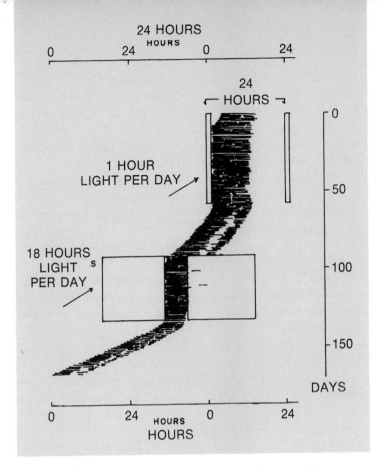

The circadian oscillation of locomotor activity in Peromyscus maniculatus *(deer mouse), freerunning and entrained. A 1:23 light-dark cycle is imposed from day 0 to day 59. The oscillation is captured into entrainment by day 6. From day 60 to day 92 the rhythm again freeruns in constant darkness. An 18:6 light-dark cycle is imposed from day 93 to day 132, when the rhythm is again allowed to freerun. Note the remarkable precision of the period of the freerunning oscillation.*

plants, and animals at all levels of complexity, including man himself. It is a general feature of the physiological organization of living things on this planet. This ubiquity is matched by the diversity of functions within the individual organism that manifest the oscillation. In the unicellular alga *Gonyaulax*, which emits light, Hastings and Sweeney have found circadian oscillations on two distinct aspects of its luminescent system: in photosynthesis, and in cell division. In mice and rats virtually every parameter studied is involved—blood chemistry, liver chemistry, cell divisions, body temperature, susceptibility to X rays and drugs, and so on. The practical importance of these more recent discoveries in mammals is obvious and immense. A given dose of the toxic substance of the human colon bacillus will kill 85% of mice treated at one point in their circadian cycle but only 5% at another (Halberg). Similar extreme effects have been reported, again by Halberg, for the heart stimulant drug ouabain.

Gertrude Stein to the contrary notwithstanding, a rose is not necessarily and unqualifiedly a rose; that is to say, it is a very different biochemical system at noon and at midnight. The phase of the organism's circadian cycle of change is a parameter of major importance the physiologist cannot ignore. Nor can the pharmacologist!

We have found, paralleling Halberg's observations, that the behavior under a fixed stimulus, and the tolerance to temperature, of the fruit fly *Drosophila*, is markedly different at different circadian phases. And along with many other laboratories we have found significant differences in the specific activity of a given enzyme system assayed in vitro (that is, in an artificial environment such as a test tube) after extraction in the middle of the subjective night. In our case we found differences in the speed of reaction between the enzyme and the substance upon which it reacts (its substrate).

The Precision, Innateness, and Temperature Compensation of Circadian Oscillations

For ease of assay, circadian rhythmicity is most conveniently studied by recording some feature of the behavior of the whole organism. In mammals, for instance, locomotion lends itself to a very useful assay. Rodents have a curious liking for exercise on running wheels. The time at which they begin and continue this activity is easily recorded by coupling the running wheel through a microswitch to an operations recorder. Every time the wheel is rotated a pen mark is made on the horizontal lines (24-hour length), as in the figure on page 307. The onset of their activity in 24-hour cycles of light and dark is precise; in nocturnal species it is at or near "sunset." When the organism is put into de Mairan's conditions of constant darkness and temperature, the period persists with remarkably clear definition, and persists indefinitely. The period of the rhythm, as measured by the intervals between onsets of activity, is circadian.

The precision and indefinite persistence of the rhythmicity are among its most striking features. The standard error of the period in some of these free-running rodents may be no more than about a minute. In other words the "error" is of the order of 1 in 1,000.

The rhythmicity is, moreover, innate to the organism; it is not learned by prior experience (in the individual) of a daily periodicity in the environment. Rigorous demonstrations of innateness have been made in unicellulars, insects and vertebrates.

Perhaps the most surprising property of circadian oscillations is the fact that their period changes only very slightly with considerable changes in temperature. This temperature compensation of the oscillation is a large topic in its own right. It has played a major role in F. A. Brown's thinking. The difficulty of giving a simple physiological explanation for it led him, in part, to his renewed concern with the possible existence and function of an external physical pacemaker (factor X). On the other hand, it was a property this writer inferred should be general if circadian oscillations were fulfilling a general clock function—a topic to which we shall return shortly.

The Cellular Basis of Circadian Oscillations

A significant result in the last ten years has been the demonstration that de Mairan's rhythms do not depend on the greater complexity of higher organisms. Single cells manifest the rhythm—in, for example, *Euglena, Paramecium,* and the alga *Acetabularia.* Sonneborn and Barnett have studied a remarkable case in *Paramecium multimicronucleatum.* "Animals" of this species oscillate from one mating type to another in the course of a single day. From what is known of the genetics of mating type in other species, this result suggests that a single gene may be undergoing a daily cycle of induction and repression.

Several workers have made attempts—mostly abortive—to try to further narrow the level of organization necessary to sustain a circadian oscillation. In particular, the question has been raised as to whether or not the nucleus of the cell, or the other material of the cell (the cytoplasm), is the site of the driving oscillation.

Sweeney and Haxo, and Richter have made some remarkable observations on the famous unicellular alga *Acetabularia,* which lends itself readily to such tasks. The chief reason that it is so useful is that parts of it can be cut away (by delicate surgical procedures) without killing the cell. The nucleus of the cell can be removed; and it is then possible to experiment with the rest of the cell contents (the cytoplasm) to see whether *it* can sustain a circadian oscillation—in this case the rhythm of photosynthesis. It can, even for as many as thirty cycles.

It is also possible to remove the nucleus from one cell and graft into it the nucleus from another cell. Schweiger *et al.* did this, making a "synthetic cell" having a nucleus whose rhythm differed by 12 hours from that of the cytoplasm. They found that the rhythm of the synthetic cell was timed by the rhythm of the nucleus.

Many attempts have been made to manipulate circadian rhythms by means of chemicals; but for the most part these attempts have been singularly unsuccessful. However, the chemicals used were not labeled with radioactive elements; the results are therefore not too significant, for we cannot be sure that the chemicals that were used entered the cells in significant amounts. But at least some of those who have attempted the work are impressed with the apparent fact that the cell rhythms are not susceptible to control by chemicals.

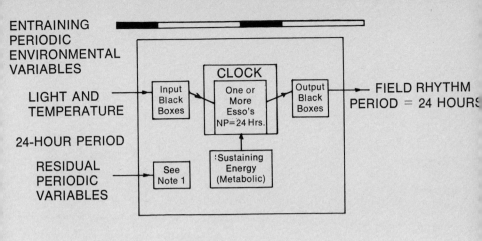

ENTRAINING
PERIODIC
ENVIRONMENTAL
VARIABLES

LIGHT AND
TEMPERATURE

24-HOUR PERIOD

RESIDUAL
PERIODIC
VARIABLES

CLOCK

Input Black Boxes

One or More Esso's NP=24 Hrs.

Output Black Boxes

FIELD RHYTHM
PERIOD = 24 HOURS

See Note 1

:Sustaining Energy (Metabolic)

Block diagram of the clock oscillator model under conditions that result in a field rhythm whose period is equal to that of the entraining (24-hour) light and temperature variables. The clock itself, upper center block, includes as a basic element one or more endogenous self-sustaining oscillators. Note: It is assumed that all other periodic variables such as pressure, humidity, air ionization, cosmic ray showers, etc., are never coupled to the oscillator; they are therefore designated residual periodic variables.

There are, to be sure, some reports of positive effects, but they are not all fully convincing. Hastings appears to have shifted the rhythm of *Gonyaulax* with cyanide arsenite and p-chloromercuribenzoate. Bünning has reported data on the effects of colchicine, urethane and ethyl alcohol on the length of a rhythm period. And Bruce and Pittendrigh found that heavy water influenced both the phase and the period of division of the *Euglena* cell. But none of these results have provided fruitful suggestions as to whether there is any chemical basis for circadian oscillations.

One of the best known studies along these lines is that of Karakashian and Hastings. Using the one-celled *Gonyaulax* they measured the effects of several drugs like those that mislead enzymes into taking them in, and then jam the mechanism of the enzymes. These are known as antimetabolites. Karakashian and Hastings knew that the antibiotic drug affects the production of messenger RNA—the primary step in the transcription of the inherited message of the cell's DNA. They found after they treated *Gonyaulax* cells with actinomycin-D that the rhythms of luminescence and photosynthesis decayed. This result is surely of considerable interest, but its meaning is somewhat uncertain. The loss of rhythmicity, as such, by no means demonstrates that the agent has affected the cell's clock; its target could as well be the coupling of the clock (or driving oscillation) to the physiological system under investigation.

The only clear demonstrations that the clock can be manipulated concern either the period of the phase, or a rhythm that is in a steady state.

Strumwasser has very recently given evidence on the action of actinomycin that promises to fulfill these requirements. In a brilliant study of a circadian rhythmicity in a single cell of a ganglion from the mollusk *Aplysia*, he has succeeded in injecting various agents into single cells.[5] The results he has published so far show a clear shift in the phase of the oscillation after the cell has been injected with actinomycin. At present, however, the rhythm cannot be measured for much more than a single cycle after treatment. We cannot, therefore, be sure that the phase-shift seen in that cycle will persist in the steady state and thus demonstrate fully that the driving oscillation has been affected and not (again) some coupling mechanism between the driver and the rhythm being assayed.

Other suggestions that the nucleic acid systems (DNA-RNA) in the cell are intimately involved in circadian rhythmicity arise from observations by Ehret and, independently, by Sweeney that ultraviolet light from a mercury lamp can reset the phase of the oscillation. An exposure of only a few minutes is all that is needed to reverse the phase completely. There are peculiarities in both Ehret's *(Paramecium)* and Sweeney's *(Gonyaulax)* results that indicate that the ultraviolet produces its effects by a route quite distinct from that of visible light.

[5] The "ganglion" is a small mass of nervous tissue that has a concentration of nerve-cell bodies which function as a coordinating center.

The Cell and Organism as an Oscillator Entrainable
by Light Cycles

In 1957 Bruce and Pittendrigh pointed out that, formally, circadian rhythms are self-sustaining oscillations, and pursued a general comparison between the relation of the organism to its environment and that of *two* oscillators, one entraining the other. In historical perspective it is clear that circadian oscillations in the cell and organism are an evolved match to the striking oscillations of the physical environment. Metabolism has evolved an oscillatory time course; and in nature that oscillation (in both metabolism and behavior) assumes a definite relation to the external cycle of physical change. That "proper" relation of phases is attained—at least principally —by the light cycle. The oscillation of a cell couples to the light cycle and is entrained or driven by it. But the light cycle, as Sachs long ago recognized, is not imposing the rhythm. Its action is strictly comparable, mathematically, to the action of one oscillation entraining another, fully independent, self-sustaining oscillator.

Entrainment implies control in two respects. The entrained oscillation assumes the period (or frequency) of the entraining cycle; when, as in the biological case, the entraining cycle is error-free, the errors inherent in the imperfections of the entrained oscillator are removed. Second, the entrained oscillation assumes a fixed relation to the phase of the entraining cycle. The net biological result is clear temporal control; specific events in the circadian cycle occur at particular times in the environmental cycle.

It is, for the most part, an act of biological faith when we go further and say that the end result is performance of given functions at the *"right time of day."* This is to follow Darwin in his confidence that some adaptive function does, ultimately, attach to such a remarkable piece of organization. To suppose otherwise is, implicitly, to appeal to something other than natural selection as the historical agent of the system.

We will not take up in detail what is now known of the

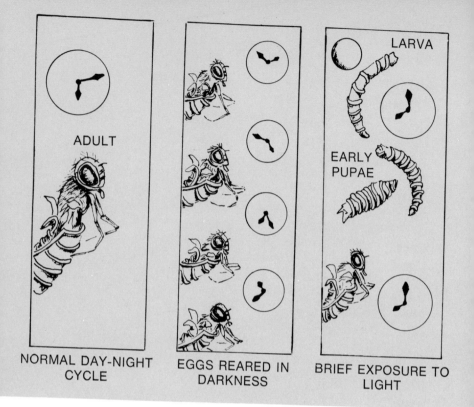

ADULT

NORMAL DAY-NIGHT
CYCLE

EGGS REARED IN
DARKNESS

LARVA

EARLY
PUPAE

BRIEF EXPOSURE TO
LIGHT

When fruit flies are reared under normal conditions of alternating light and darkness, the adults emerge from their pupal cases near dawn, left. When reared in continuous darkness, the flies eventually lose their time sense, and emerge at random times, center. But a brief exposure to light during early development resets the clock, so that the adults then emerge at whatever time the exposure was given.

mechanism of entrainment by light. It will be enough to note a few points of general interest. First, light cycles are universally effective in entraining circadian oscillations. In cold-blooded animals temperature cycles can also entrain, but they are probably less powerful agents; and recent work by Mr. Zimmerman in our laboratory shows that this is certainly the case with fruit flies. Second, observations by Halberg, Richter and my laboratory have shown that in mammals the light, in its entraining function, is transduced by the eye; blinded mice, rats and hamsters fail to entrain to light cycles.[6] It is, however, likely that this route is historically secondary. If, as is surely true, the hypothalamus (the region of the forebrain which controls various centers which in turn control visceral activities, water balance, temperature, sleep, and so on) acts as a driving center in the system of circadian oscillations within a vertebrate, its coupling to the light cycle of the environment will almost necessarily demand an intermediate coupling to a superficial photoreceptor. (It is noted, however, that Ganong's remarkable demonstration that visible light can penetrate into the brain stem of vertebrates takes some force from this argument.) However, it is equally clear, as experimental fact, that the circadian system of the fruit fly can be entrained when the fly is in the larval stage, and has no organized photoreceptor. And several workers (including Lees and Williams) have shown that photoperiodic induction (which, I shall argue later, is a function of the circadian oscillation) can be effected by the action of light absorbed directly by central nervous tissue.[7] In *Paramecium*, other unicellulars, and in green plants, the question of an organized "eye" does not arise. They, too, are entrainable by light. The general conclusion is that some molecule in the cell, not specifically devoted to receiving light in the usual visual sense, absorbs the entraining light and is intimately connected with the driving mechanism of the circadian oscillation.

The third point I wish to emphasize about light cycles as entraining agents for circadian oscillations is the general result that the photoperiod in each cycle—the fraction of the period occupied by light—has a major effect on the phase, and also the waveform of the entrained rhythm (insofar as one can measure or infer this).

---

[6] A "transducer" is a device (in this case the eye) which can be excited by one or more transmission systems, and which in turn excites one or more other systems.

[7] An "inductor" does something to cells that decides—out of a number of possible types of development—which one will actually be realized.

Chronometry by Circadian Oscillations: Celestial Orientation and Photoperiodism

Victor G. Bruce and Pittendrigh have suggested that the resurge of interest in circadian rhythms since 1950 largely derives from the remarkable studies by Gustav Kramer and Karl von Frisch, which showed that birds and bees can maintain a given direction throughout the day using the sun as a compass. They compensate for the movement of the sun with the aid of an internal 24-hour clock. The experiments supporting these remarkable conclusions are classics of experimental zoology. Hoffman and others have shown that the animals' clock is phased to local time by being coupled to the light cycle of the environment. In starlings Hoffman has shown further that the clock will continue to operate in continuous dim light, and proves to be—in these free-running conditions—a circadian oscillation. Its free-running period is about 23½ hours.

Kramer's initial demonstration prompted my own reinvestigation of the temperature relations of the circadian system in the fruit fly *Drosophila*. On the hypothesis—then—that circadian oscillations were the evolutionary foundation of Kramer's clock, it seemed to me that to be useful in this respect they should be so compensated as to be essentially independent of temperature; and this proved to be so. Since then (1954) temperature compensation has been shown to be a universal feature of circadian oscillations even in single cells.[8] And since 1950 time-compensated sun-orientation has been discovered in a remarkably diverse array of higher animals.

It is clear, however, that this spectacular clock function is recent in the history of life; it has exploited already existing circadian oscillations; it does not account for their initial evolution.

Nearly fifteen years before the work of Kramer and von

[8] Victor G. Bruce and Colin S. Pittendrigh: "Temperature Independence in a Unicellular 'Clock,'" *Proceedings of the National Academy of Sciences, Washington, D.C.*, Vol. XLII (1956), pp. 676–81.

Frisch, Erwin Bünning had related circadian rhythms to a quite distinct set of phenomena—those of photoperiodism. Garner and Allard showed, in 1920, that the switch from vegetative to floral growth in some plants was controlled by the number of hours of daylight—the photoperiod—in each daily cycle. Bünning's suggestion in 1936 was that the endogenous "daily" rhythmicity of plants was causally related to this control. He saw what we now call the plant's circadian rhythm as consisting of two half cycles—one light-loving (which he called "photophilic"), and the other dark-loving (which he called "scotophilic"). The latter, in its usual phase relation to the environmental day, lies in the nightly dark period. (In other words, photophil is equivalent to the day phase and scotophil the night phase of a circadian rhythm). He suggested that as the length of day changed, the early scotophil would be illuminated or not, according to the season. When the early scotophil was illuminated in long-day plants, the switch to the flowering stage was closed; in short-day plants illumination of the early scotophil kept the switch open. His hypothesis, translated into the current jargon, was, in fact, that the circadian oscillation of the plant was serving as the clock that effects the time-measurement implicit in photoperiodism; it was the clock measuring the length of the daily photoperiod.

That hypothesis was a brilliant stroke in its day. It foresaw the current emphasis on time measurement as such as the knottiest problem in photoperiodism, and it foresaw the current treatment of circadian oscillations as biological clocks in general. Yet the hypothesis has met with stubborn opposition by students of both plant and animal photoperiodism. That opposition is now weakening, and the evidence today leaves essentially no doubt that Bünning's proposition is basically correct.[9]

---

[9] Colin S. Pittendrigh and D. H. Minis: "The Entrainment of Circadian Oscillations by Light and Their Role as Photoperiodic Clocks," *American Naturalist*, Vol. XC (1964), pp. 261–94. See also Colin S. Pittendrigh: "On the Mechanism of the Entrainment of a Circadian Rhythm by Light Cycles," in Jürgen Aschoff, ed.: *Circadian Clocks* (Amsterdam: North-Holland; 1965), pp. 277–97.

On the Ancient, or Primary, Function of Circadian Oscillations

Time-compensated orientation by the sun is surely a recent development in the history of organisms. And it seems likely that classical photoperiodism is of more recent origin than circadian oscillations in general. Bünning (1957) has, to be sure, suggested that their role in photoperiodism is their primitive function—the source of the original selection pressures that generated them. That seems, however, very unlikely to this writer. Circadian oscillations are widespread in one-celled organisms, for instance, in which with one recent exception (Steele, 1965) there is no evidence of seasonal control by photoperiod. Bünning's attempt to find functional significance for them in photoperiodism again emphasizes a point already noted.

Since Darwin's day botanists have been hard-pressed to find adaptive meaning for the particular manifestation of circadian rhythmicity they most often meet and study, namely, the "sleep" movements of leaves. I granted earlier in this paper that the proposition of a "right time of day" for a given metabolic function is in large part an act of faith. The fact is that, for many of the circadian rhythms selected as obvious and easily assayed, there is no very well defined adaptive function. Remmert (1962) has justifiably questioned—as at least not proven—the adaptive use I suggested some years ago (1954, 1958) for the rhythm of hatching of fruit flies. Indeed, Remmert is perplexed about the adaptive function of a huge number of rhythms of metamorphosis in insects.

The sheer diversity of activities in organisms that manifest circadian rhythms raises a clear question. Are they of independent evolutionary origin, or have they a common historical origin? Of course, there is no doubt that many of the details of diverse circadian systems are of independent origin, and that some of their formal properties owe their similarity to convergence, that is, the independent evolution—in organisms of widely different ancestry—of similar structures and similar functions. It is difficult, however, to accept a convergence from independent origins as the explanation of all the similarities they show, especially those most "improbable" physiological features: (1) precision, and (2) temperature compensation. They are all similar, too, in their entrainability by light. It is difficult, in short, not to retain as a working position the view that there is a common core to the structure of circadian oscillations that is ancient; that they evolved to serve a function not yet explicitly recognized; and that the few clearly defined functions they serve today are secondary exploitations of an organization that arose to meet other immediate problems.

If such an ancient, primary function indeed exists, it is certainly not known, and the discussion could well rest there. I believe, however, that recognition of the issue is itself important. The analysis and explanation of the functions of organisms is not the only nor even the main task of the biologist; but even when he is primarily concerned, as I am, with physiological explanation, he cannot afford to neglect functional issues as touchstones to progress. Living organization is the product of an historical development molded by natural selection whose only concern is in fact functional. It is then possible that by recognizing the possibility of a primary function, so far not recognized, that we may be led to useful new avenues of question and analysis of causes.

The Light Cycle as the Primary Agent of Selection

The relation of the light cycle to circadian rhythmicity is today recognized only as that of its entraining agent, but it is a reasonable speculation that the daily alternation of light and darkness was the historical cause (that is, the selective agent) of circadian oscillations in the first place.

This line of thought comes from recalling the requirements for organization in a chemical system. The chief of these is that the biochemical reactions cannot proceed spontaneously at the prevailing levels of free energy, that is, at the prevailing capacity of the system to perform work. Thermochemically this means, of course, that the reactions the cell employs have energy barriers that cannot be broken at prevailing temperatures; they proceed only on command which rests, first, with enzymes, and finally with the store of information in the nucleus. Little attention seems to have been given—in this general context—to the problem of visible light as a source of energy that threatens organization.

Of course, the fact is that most of the cell's constituents are colorless; and thus visible light cannot cause uncontrolled activation. It may well be that in the history of the cell there has been selection for colorless molecules, but if that is true (and it seems likely) it is a fact that, for some functions, colorless molecular devices have not been found. The yellow pigment of cells, along with the enzymes which mediate the use of oxygen, are examples of ubiquitous and important molecules that are colored—yet where color has no known function.

No attention seems to have been given to the consequences of illuminating those molecules whose color has no obvious function. At any rate it is surely reasonable to consider, at least, the likelihood that some subroutines in the cell's overall tasks are impaired when the molecular piece-parts are activated in the flood of visible light that falls on the cell every day. To that extent the photochemical threat to organization could be escaped if some chemical activity were regularly carried out at night.

A miscellany of otherwise disconnected facts has suggested to me that we should not ignore this line of thought, and further that the activities may concern the cell's central controls—those of the synthesis and specification of proteins. I noted earlier that there is suggestive but not compelling evidence from treatments with actinomycin and ultraviolet that the nucleic acids are intimately involved in the cell's driving oscillation. There is, too, the long known fact that an enzyme system having to do with repairing ultraviolet damage to the genetic material absorbs visible light; this is known as photoreactivation. The enzyme seems to be concerned with the mechanism of genetic recombination—at least in bacteria. And the miscellany is completed by the observation of Sulkowski, Slonimski and colleagues that gene induction (in yeast) can be inhibited, at least for some hours, by visible light, during the transition from anaerobic to aerobic metabolism.

It is certain, at least, that part, and perhaps an important part, of the central control mechanism absorbs visible light, and that its functions are significantly affected by the activation that results. The possibility arises that one way of coping with this activation is to restrict the steps—perhaps gene induction itself—to the daily dark period.

Circadian Oscillations as Synchronizing Gates in the Timing of Developmental Steps

I noted some years ago (1954) that the circadian rhythm in the fruit fly was acting like a gate that synchronizes the hatching of the adult fly from its pupal case. In spite of Harker's (1965) recent discussions, this remains clearly true. But recent work by Skopik and me at Princeton shows that the synchronization manifest at this act of eclosion must occur much earlier in the development of the pupae. Harker's data, which she interprets radically differently, show, in our view, the same result. When a circadian oscillation starts in these flies in any stage of its development, the rest of the development takes a time to finish that is strictly in keeping with the period of the oscillation.

We are currently concerned with the possibility that new subroutines in development are gated by the circadian oscillations in the system, in a manner broadly comparable to that of a master clock in a synchronous computer. In such computers the inputs necessary from a variety of subroutines— some slow, some fast—are guaranteed to be available by postponing the start of the next round until a gate is opened; and that gate, timed by a master clock, opens with a frequency adjusted to the slowest subroutines. Synchronous computers are, thus, slower than the more elaborate devices that are not synchronous, but they are cheaper. All real computers involve a mix of synchronous and asynchronous features. And it is obvious that the organized reading and execution of the cell's DNA is not always regulated by synchronous gates— and certainly not by gates that open or close with a period as long as 24 hours. But selection may have seized the opportunity of circadian oscillations, especially if they involve the reading of the message, to buy a degree of organization more cheaply than is otherwise possible. There is an obvious appeal to the idea of a temperature-compensated oscillation functioning as a synchronizing device in a system whose piece-parts are conspicuously dependent on temperature. Fruit fly eggs raised at 10° and 28° Centigrade yield flies that are essentially the same; the development of an organism's form and features— and its development of specialized cells—are compensated for temperature; the reading of the message is compensated for temperature also.

It is clear that if the cell of most organisms has restricted the light-sensitive step to the daily dark period, it has also by that very fact set up a synchronizing gate that is compensated for temperature. What we are now pursuing in the fruit fly is precisely this idea. The working position is that the hatching of the adult fly is timed as by the opening of a gate; the fly may actually be ready to hatch before the gate opens, but it cannot do so before then; on the other hand, if it should be ready to hatch only minutes after the gate closes, it must then wait through the next cycle until the gate opens again. This is regarded as only a special case of a more general phenomenon —the gating by the circadian oscillation of the induction of genes. The factor responsible for the induction may have been available earlier, but the actual induction awaits the scan of the message, and that takes place only once a day. We are encouraged in this view by Harker's facts, which she interprets in a fundamentally different way. She seems almost to deny the existence of oscillations in individual flies, and regards the time of hatching as determined by adding up the intervals between earlier steps of development. She ignores the fact that the times of development are always some multiple of the period of circadian oscillation; and further, that her own data indicate that each new step, like the appearance of color in the eye or the wing, tends to be fixed to the phase of the light cycle—and hence to the circadian oscillation.

This excursion into the history and primary functional significance is frankly speculative, and would be unjustified if it did not suggest new experimental work. Fortunately, it does. We are, in fact, greatly encouraged in testing the idea that events that occur in the middle of development can be timed by circadian oscillations as well as the final event of the hatching of the adult; and that oscillation can be manipulated by brief flashes of light.

It is not hard to relate this line of thought to the idea that circadian rhythms are involved in photoperiodic induction. The change in day length that occurs with season effects a switch in the program of metabolism—as from vegetative to floral growth. That change calls forth a part of the message of the nucleus that would otherwise be ignored or suppressed. In short, photoperiodic induction must, in last analysis, involve induction by specific genes. The prospect that induction is restricted to a daily scan of the message in each dark period is clearly compatible with the further idea that induction of a given gene lying toward the end of a scanning sequence is as sensitive to light as the inductions Sulkowski *et al.* have observed in yeast cells. There is, of course, evidence now from studies of bacteria that any scan of the message—for replication or for reading—will proceed along a line down the fixed sequence of the genetic code that is specific to the species. The meager facts available indicate a scanning time for replication (in cells having complex nuclei) of 6 or 8 hours; it is reasonable to anticipate that a scan for reading will take about the same time—which is in the right range for measurement of the length of the night.

The scope of the speculation I have ventured here guarantees that much of it will be wrong. Its justification is in part its testability, as I noted. But it is also justified if, in asking whether there is a primary function for circadian oscillations that is not yet identified, the speculation prompts new questions and experiments.

18.   Ehret: Clocks at the Base of Life

Twenty-five miles southwest of Chicago, in the heart of a thousand acres of fields and woodlands, lies a research scientist's dream—the Argonne National Laboratory of the United States Atomic Energy Commission. There, Charles F. Ehret and his associates in the Division of Biological and Medical Research are exploring the individual molecules within the nuclei of single animal cells in their search for the mechanism of the living clock.

In less than thirty years the Division of Biological and Medical Research at Argonne has grown from a small but essential advisory service in a South-Side Chicago squash court to one of the premier scientific institutions in the world.

Its history began with the establishment of the Manhattan District's Metallurgical Laboratory on the brink of

World War II. Acting under emergency orders, Arthur Holly Compton, professor of physics at the University of Chicago, shouldered the awful responsibility of directing the project that achieved man's first controlled release of nuclear energy.

In spite of his top defense priorities, however, the only space that Compton could find for his team of physicists was in a squash court beneath the west stands of Stagg Field, then unused because of President Hutchins's order banishing football from the University of Chicago. The squash court was some 30 feet wide, 60 feet long, and 25 feet high—a niggardly allotment of space in which to construct an atomic pile. Still, Compton felt that with the team of outstanding scientists that had been made available to him, "the impossible would take just a little longer."

Leader of the team was Enrico Fermi, who had only recently arrived from Italy, and who even then was a Nobel laureate in physics of some four years' standing. Other names to conjure with were those of Herbert Anderson, George Weil, Leona Wood, and Walter Zinn. Fermi had thought that the pile could be made to work by May or June of 1942, but it was not until the afternoon of December 2 that the pile actually became operational. A controlled chain reaction was maintained for twenty-eight minutes. Then at Fermi's order George Weil pushed the cadmium master control rod back into the pile, and the chain reaction subsided.

Quietly Arthur Compton left the squash court. The son of a Presbyterian minister, and himself a man of deeply religious convictions, Compton had undertaken the atomic work with the greatest reluctance, and then only because of his belief that nothing short of an atomic bomb could bring a quick end to the war. Weary, and torn between feelings of triumph and foreboding, he placed his now-famous long distance call to James B. Conant in the Office of Scientific Research and Development at Harvard.

"You'll be interested to know that the Italian Navigator has just landed in the New World."

"Is that so? Were the natives friendly?"

"Everyone landed safe and happy."[1]

A great deal of the credit for the safe landing belongs to Enrico Fermi for the meticulous care with which he made his theoretical calculations and the precautions that he took against every foreseeable contingency. Nevertheless, this new world was to a great extent unknown, and unforeseeable hazards lurked there.

For just that reason expert radiobiologists and physicians followed each new experiment step by step, and counseled extraordinary precautions. At stake was the safety—even the lives—of scientists who were exposed to larger numbers of neutrons than man had ever before experienced, and who were working with radioactive substances whose effects on human beings were still largely unknown. Small wonder that these medical men looked forward to a day when the pressures of war might lessen, and time and facilities might become available for the gathering of knowledge in this uncharted realm.

Nothing could have pleased them more than to see the medical and biological research facilities at Argonne today. Elite groups are studying the biochemistry, biophysics, and general biology of reactions at the molecular level; of developmental cell biology; of genetics; of plant and animal physiology; of general radiation biology; and of theoretical and analytical biology. In support of this activity is one of the largest animal facilities in the country, where pioneering research on animal care is in progress.

On the thousand-acre surrounding site of natural fields, marshlands, and woodlands, wildlife abounds, so that today's pressing problems in ecology are being attacked in complementary laboratory and field research.

For the fundamental laboratory studies, the most advanced and sophisticated equipment is available for research on the organelles (the specialized structures within cells, such as the

---

[1] Laura Fermi: *The Story of Atomic Energy* (New York: Random House, Inc.; 1961), p. 89.

mitochondria, the Golgi complex, the ribosomes, the contractile vacuole, and so on). And of prime importance is the opportunity for free interchange of ideas among experts in many different disciplines, both on the Argonne staff itself, and with the faculty members of the thirty-three universities that make up the cooperating Associated Midwest Universities.

To this wellspring of opportunity went Charles Ehret in 1948, after receiving his Master's degree in zoology from the University of Notre Dame. He followed the fellowship program of the Associated Midwest Universities, under which he did his doctoral research at Argonne, and met his other graduate requirements at Notre Dame. He was awarded his Ph.D. in 1951.

His research from the very beginning was remarkable for its singleness of purpose: to seek a deeper and deeper understanding of the events—virtually limitless in kind and number —that occur within single living cells. And of course at Argonne he had unparallelled facilities for pursuing that objective. But he did not permit his singleness of purpose to limit his interests. On the contrary, these have been very broad, and he has made a point of bringing to his own field fresh ideas from a surprisingly wide variety of sources.

As cross-breeding of animals and plants often leads to much improved varieties, so it is with ideas also. The *Random House Dictionary* defines "creative imagination" as "recombining former experiences in the creation of new images directed at a specific goal or aiding in the solution of problems." Charles Ehret's work provides a fascinating example of how this process can operate.

Very early in his study of cell physiology Ehret became familiar with the accurate rhythms of cells, and he also knew that most biologists were beginning to think that the ability to measure time was an essential requirement for an organism to survive; to maintain life in a world geared to the passage of time, a plant or an animal had to know what time it was.

Accepting this, Ehret reasoned that it ought to be possible to imagine some form that a living clock might take, and some principle by which it might operate.

How would one imagine such a form and such a principle?

Ehret thought that a logical approach would be to make a study of all the different kinds of devices that man had used through the ages to measure time, and then see if any of them might resemble the kind of clock that a cell could possess. That search naturally led him far afield.

He found, for example, that from the tenth to the eighth century B.C. the Egyptians used a T-shaped bar whose shadow, cast by the sun, served as a primitive kind of sundial. As examples of later mechanical oscillators, he studied the Greek water clock and Galileo's pendulum clock (Chapter V). And to bring himself right up to date, he studied the atomic clock, which is regulated by the vibrations of cesium atoms, and loses no more than one second every three thousand years. For amusement, he assembled his own collection of all kinds of timekeepers: hourglasses, pendulums, old-fashioned mechanical alarms, flower clocks, weight-driven verge escapement clocks, and even a clock in which dots of white mold growing on pink agar in a glass petri dish told the time.

But the clock that kept demanding his attention, and that he returned to over and over, was a modern model of King Alfred's rope clock—a yellow candle cast as a spiral, and marked off by horizontal brown bands, each an hour's burning distance apart. King Alfred's original, according to a ninth-century monk, was made of lengths of rope twelve inches long that had been saturated with some three quarters of a pound of a mixture of beeswax and tallow. These lengths burned at quite constant rates of three inches an hour, and thus lasted for four hours. Marked off into twelve equal divisions, the time of burning from mark to mark was twenty minutes. And of course as soon as one length burned out, the next was lit, and the burning of six lengths spanned a day.[2] In Ehret's mind, King Alfred's clock began to take a dim shape as the first "image" in an act of creative imagination.

---

[2] Edward Wenham: *Old Clocks for Modern Use* (London: Paul Hamlyn, Ltd.: Spring Books; 1965), p. 13.

The second "image," as it happened, also originated in England, and also was in the form of a spiral. This was the famed double helix—the DNA spiral—whose structure was worked out by James D. Watson and Francis H. C. Crick in 1952, and for which they and M. H. F. Wilkins were awarded the 1962 Nobel Prize in medicine. A monumental amount of evidence has since confirmed the fact that DNA carries life's genetic code: that it functions as the template upon which all living organisms build up their complicated organic molecules that differentiate them as species from other species, and as individuals from other individuals within the species.[3]

Ehret therefore now had before him two images that were striking in their similarities, but also striking in their differences. He had, in a word, two things that were ideally suited to examine through the process of reasoning by analogy. This kind of reasoning is so powerful and so productive in scientific research that we may well review it briefly here. A particularly apt description of its operation has been given by Monroe C. Beardsley of Swarthmore College:

"First, of course, an analogy is a comparison: we say that two things have certain characteristics in common. But we do not generally call a similarity an analogy unless it is somewhat complex. Two white things are similar, but not analogous. But if two things have several characteristics in common, and especially if these characteristics are systematically related to each other, then there is a structural similarity between the two things. A map has an analogy to the territory it maps, because the order of points on the map corresponds to the order of cities in the territory. . . . Finally, we do not usually speak of even a complex similarity as an analogy unless the two things we are comparing belong to basically different categories of things. Two Wyoming road maps are extremely similar, but they are not analogues; two eggs may be indistinguishable, but they are not analogues of each other. The map and its terri-

[3] For an absorbing account of the role of DNA in the development of the life sciences, see Ruth Moore: *The Coil of Life* (New York: Alfred A. Knopf, Inc.; 1962).

tory, however, are very different kinds of things; the former lacks certain essential characteristics of the latter. So their likeness, striking as it is, reaches across a deep contrast in kind. . . . To discover so close a similarity *despite* so great a disparity is to discover something unexpected and impressive."[4]

Clearly, then, King Alfred's clock and the DNA double helix were analogues. However, to make the most profitable use of an analogy, one must examine the similarities and differences between the two things in detail. Ehret now proceeded to do this, using for convenience his modern candle model rather than King Alfred's waxed rope.

The differences between Ehret's two things were, of course, more striking than the similarities. The candle was a dead object, made by man; the DNA helix was a very large molecule that was a part of a living cell, and that had been made during the life process from a matching helix. The candle was perhaps an inch in diameter and one foot high; the DNA molecule in the nucleus of, say, a human cell, would be about three feet long if it could be stretched out straight, but it is submicroscopically thin; and in chromosomes of the cell, it is packaged so tightly that it is only a few hundred-thousandths of an inch across. Or, to use a figure, if it were stretched out straight and imagined to have the one-inch diameter of the candle, it would be five miles long. Again, when the candle has burned to the socket, it is done with; DNA continues to replicate itself throughout the life of the cell.

The similarities between the candle and DNA were these: Both were shaped as spirals, and both had patterns that repeated themselves from one end to the other. The candle had an inch of yellow wax, and then a thin horizontal brown band, repeated over and over; DNA had a repeating pattern also, but far more complicated—a recurring sequence of adenine, guanine, cytosine, and thymine.

Here Ehret reached the critical point in his reasoning process.

---

[4] Monroe C. Beardsley: *Thinking Straight*, 3rd edn. (Englewood Cliffs, N.J.: Prentice-Hall, Inc.; 1966), p. 131.

Just here, as he carried out his analogy one step further, the crucial question leaped to his mind. The candle was used to measure time! Was DNA doing the same thing? Could a cell use DNA in any analogous way to measure time?

A flood of questions followed. Might the repetitive pattern of the DNA act like the escapement of a fine watch? What would be the role of messenger-RNA in the process?[5] How would the associated process of the synthesis of proteins from amino acids relate to timekeeping? Most important, what particular organism could Ehret use in experiments designed to answer such questions; ideally it would be the simplest one-celled organism known that could tell time.

But wait!

Had Ehret really asked himself the right question in the first place? He had better be more confident of that before he rushed to the laboratory and started a furious barrage of experiments. It was no easy matter for Charles Ehret to hold himself in check, for he is an intense man, motivated by a powerful drive to get things done. But he is also a highly intelligent one, and he knew that he could make the surest progress if he kept his feet planted firmly on the ground of facts. Calmly then, he sat back and reviewed the background of the whole problem.

What *were* the established facts?

As fundamental background for his thinking, Ehret had long since reviewed the various classification schemes that had been used to divide living things into groups. Of course there was the classical division between the plant and animal kingdoms, and the further division of these into phyla, orders, classes, and so on. Then, by the 1950's, microbiologists had come to recognize that from an evolutionary standpoint, micro-organisms could be divided more logically into three kingdoms,

---

[5] Messenger-RNA is a particular kind of ribonucleic acid which is synthesized in the nucleus and passes to the ribosomes in the cytoplasm; it combines with RNA in the ribosomes and provides a template for the synthesis of an enzyme or some other specific protein. (Claude A. Villee: *Biology* [Philadelphia: W.B. Saunders Company; 1967], p. 696.)

and to the traditional plants and animals they added a kingdom of protists—from the Greek "protistos," meaning the very first.

Having set up this third kingdom, the next step was to find a logical basis for classifying its members, and, since the new kingdom was limited to microorganisms, the obvious place to look for distinguishing features was at the molecular level. Microbiologists were not long in finding that. It lay in the character of the cell nucleus, and made possible the division of the new kingdom of protists into just two large groups. All members of the first group had within the cell a nucleus that could be clearly identified, and that was very complex; these were called the higher protists. All members of the second group had far simpler nuclei, often difficult to identify, and often lacking in many of the features found in the higher protists; these were the lower protists.

Members of the higher protists included the protozoa, the fungi, and many of the algae. Lower protists included the bacteria and the blue-green algae. Looking again to the Greek for biological terms, the nomenclature specialists called the higher protists "eukaryotes," and the lower protists "prokaryotes."

Finally, it was already well established that all higher plants and animals had cell nuclei that were easily identifiable and very complex. Therefore, on the basis of this classification scheme, the higher protists among the microorganisms, together with *all* higher plants and animals fell in the same grand class—the eukaryotes. If, then, Ehret could find the clock in the simplest eukaryote, that could well lead to the identification of *all* living clocks!

Here was a straight line of common cell characteristics that Ehret could use to run right down the evolutionary scale to the simplest eukaryote that could tell time. He could start either from man and go down the animal scale to the simplest cell, or he could start from the most complex plant, and go down the plant scale to the simplest cell. From either starting point he should arrive at the same place—the simplest eukaryotic protist known to science. That turned out to be *Paramecium*.

He tells some of the reasons for this choice in a seminar tape that was prepared at Argonne for college and university audiences:

"The smallest and simplest animal cell in which, to date, circadian rhythms have been clearly shown is the ciliated protozoan *Paramecium*. A conspicuous rhythmical parameter of this genus is its capacity to mate. Mating normally occurs at daytime, not at night. Under constant conditions in darkness, the mating capacity rhythm continues for as long as a week, with a circadian period of about 22–23 hours that is nearly independent of the temperature over the range from 17°–30°C. *Paramecium's* biological clock is readily and predictably reset by radiant energy. In chlorella-less[6] *Paramecium busaria* the action spectrum for resetting the phase includes peaks in the near ultraviolet, the blue, and the red regions of the spectrum, and *no* reversals by far-red wavelengths have been observed, suggesting the involvement of porphyrin-like but not phytochrome-like pigments. Of even greater interest was the discovery that not only is the *Paramecium* clock resettable with high efficiency by far ultraviolet radiant energy but that this effect is itself photoreversed by white light. This finding suggested that the subcellular clockworks reside within the framework of the control systems for nucleic acid metabolism in the non-dividing cell. Consistent with this hypothesis were the subsequent experimental observations of significant oscillations of nucleic acid (NRA) and of nucleotide-coenzyme fractions of the cells that correlated directly with the previously established physiological measures of circadian activity."[7]

For a more general audience, Ehret would have gone on to say that ultraviolet light damages the DNA helix, but the cell can repair that damage if it is subsequently exposed to white light. This is what—in its simplest terms—led Ehret to con-

---

[6] Under natural conditions the *Paramecium* cell usually has symbiotic cells of the green alga *Chlorella* inside it. As Ehret points out, his cultures were free of *Chlorella*.

[7] Charles F. Ehret: "The Sense of Time: Its Radiobiological Implications and Consequences" (Office of College and University Cooperation: Argonne National Laboratory, Argonne, Ill.; May, 1966).

clude that the clock mechanism must be associated with the control system for nucleic acid metabolism.

Further evidence pointing in the same direction had also been accumulating in other laboratories, in particular, studies of the effects of actinomycin-D on microorganisms. Actinomycin-D is an antibiotic derived from soil bacteria, and it is known to inhibit the synthesis of DNA in the cell. Hastings and his coworkers had already shown that actinomycin-D stops the *Gonyaulax* clock (Chapter XIV). Sweeney had been able to reset the *Gonyaulax* clock with ultraviolet light. McMurray had shown that under the influence of substances that inhibit protein synthesis, the four rhythms of Gonyaulax cannot be uncoupled from each other by phase shifting. Thus it seemed to Ehret more and more likely that the best clues to the location of the clock pointed right to the cell, and within the cell, to the base complex of life—DNA, and messenger-RNA, and the associated protein synthesis reactions.[8]

Had the complexities involved now become so great that the analogy with King Alfred's clock might collapse? Charles Ehret may have wondered, but he did not change his course. He enlisted the cooperation of two Argonne associates, John J. Wille, a specialist in the reactions of very large organic molecules, and Ernesto Trucco, a biophysical mathematician, and together they renewed the attack on the problem.

By 1967 they felt that they had made enough progress that they should publish the results of their thinking, and learn what the reactions of other biologists might be. Admitting right at the outset that they were giving the barest outline of an idea, which for the time being remained undeveloped, they offered a "phenomenological model for biological circadian timekeeping."[9]

---

[8] Charles F. Ehret and John J. Wille: "The Photobiology of Circadian Rhythms in Protozoa and Other Eukaryotic Microorganisms," *Photobiology of Microorganisms*, ed. Per Halldal (New York: John Wiley and Sons, Inc.; 1970), chapter 13.
[9] Charles F. Ehret and Ernesto Trucco: "Molecular Models for the Circadian Clock. I. The Chronon Concept," *Journal of Theoretical Biology*, Vol. XV (1967), 240–62.

The "chronon" concept which they had arrived at is shown in the accompanying figure. Obviously the whole process is extraordinarily complicated; no attempt will be made here to discuss it in detail, for that would be quite beyond the scope of this book. However, by great oversimplification of the model, we can arrive at an approximation of Ehret's thinking. He visualizes very long DNA molecules at the heart of the time-keeping sequence, and refers to them as chronons. Upon the individual strands of the DNA helix, messenger-RNA is built up until it fits the full length of one DNA strand. Simultaneously a series of associated chemical reactions takes place in sequence, and the rates of these reactions may be thought of collectively as functioning like the escapement of a watch. The entire sequence thus serves as a precise timekeeping mechanism that is independent of temperature within quite wide limits.

In spite of the complexity of the model, Ehret thinks of it as showing "just the bare bones . . . , omitting all complexities that are not absolutely essential." If that is the bare bones, then the figure indeed serves to dramatize the staggering complexities that face molecular biologists today.

To cope with such concepts obviously requires great intellectual force and stamina. It also places heavier demands on physical strength and endurance than most people realize. Fortunately, Charles Ehret is abundantly endowed with both. His intense vitality spills over in a profuse display of gestures and mannerisms. He is never still. When he speaks to an audience about the "eukaryotic-circadian principle" he uses no notes, and in his directness of address he compels his audience to relive his laboratory work right along with him.

Completely free of self-consciousness, he brushes back wayward strands of reddish-brown hair, raises his eyebrows high, and fixes his listeners with intent stares as if to bring about direct thought transfer even more effective than his rapid-fire delivery.

When he is done, he stops. Then he flashes onto the screen some hilariously funny slide. Laughing as they leave, his listeners realize that beneath a brilliant display of erudition

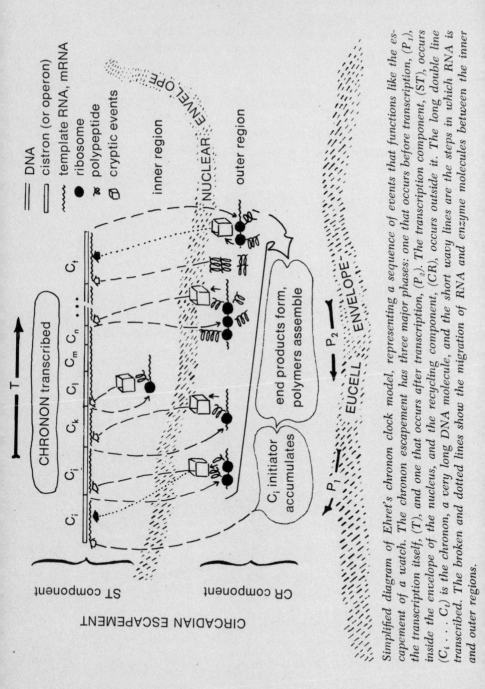

Simplified diagram of Ehret's chronon clock model, representing a sequence of events that functions like the escapement of a watch. The chronon escapement has three major phases: one that occurs before transcription, (P_1), the transcription itself, (T), and one that occurs after transcription, (P_2). The transcription component, (ST), occurs inside the envelope of the nucleus, and the recycling component, (CR), occurs outside it. The long double line ($C_i \ldots C_t$) is the chronon, a very long DNA molecule, and the short wavy lines are the steps in which RNA is transcribed. The broken and dotted lines show the migration of RNA and enzyme molecules between the inner and outer regions.

they have caught a glimpse of a very human man. Those who take a moment to check his biography in *American Men of Science* learn that he is married and the father of seven children.

When Ehret refers to the chronon concept as just the bare bones he is surely guilty of undue scientific modesty. At the least, he has created a rare product of the creative imagination that has already given deeper insight into the formidable problem of how living clocks work. Many keen students of biorhythms think that he has also provided a sound structural framework that will, in days to come, be fleshed out by the sinews of a living theory.

"*In any case, I wish
to emphasize that our ignorance
of living organization,
of how living matter works,
is in general so profound, and
our lack of real theory
(in the sense that the physicist
uses that term) so complete,
that we cannot afford the luxury,
or rather the scientific arrogance,
of insisting that we know
the space environment
will bring us no surprises.
In short, we cannot
afford an arrogance that may
cause us to lose the possibility
of major discovery.*"
COLIN S. PITTENDRIGH,
1963[1]

19.   Lindberg: The Pocket Mousetronauts

Dr. Robert G. Lindberg, head of the Bioastronautics Laboratory of the Northrop Space Laboratories in Hawthorne, California, is planning for the day when he can send a tiny desert pocket mouse into deep space. He believes that it is critically important for space explorers to know how living clocks will behave when they are removed from the fields of all geophysical forces that surround the earth. The very survival of the astronauts who will probe deeper and deeper into space will depend on a knowledge of how their own clocks will react to this environment. This knowledge should come from experiments on lower organisms—not from premature experiments on man.

[1] Colin S. Pittendrigh: "Scientists' Testimony on Space Goals," U.S., 88th Congress, House Committee on Aeronautical and Space Sciences, Hearings of June 10–11, 1963 (Washington: U.S. Government Printing Office; 1963), p. 75.

"If the circadian rhythms of man are in any way coupled with terrestrial cues," Lindberg writes, "the probability of his satisfactory performance on prolonged space missions will be low. Within this context, studies of circadian rhythms in deep space can have as high a priority as the required measurement of ionizing radiation in space."[2]

Short of a direct penetration into deep space, such as a mission to Mars, there are three orbital missions that would provide exceedingly useful information if a small automated biological laboratory were aboard: an orbit around the earth, an orbit around the moon, and an orbit around the sun. While the first two would be the most readily accomplished, vehicles in these orbits, at least in theory, would not be completely free from the influences of the earth's fields. A satellite in orbit two hundred or three hundred miles above the earth's surface could still be argued as being within the influence of "pervasive geophysical forces" that animals might sense. A satellite in orbit around the moon would approach and retreat from the earth in rhythmic cycles, and thus organisms aboard such a craft might be entrained to this rhythm; if entrainment from that cause should occur, the rhythms under investigation would be masked. Both of these effects would be vanishingly small in a heliocentric orbit, and that is the orbit that Lindberg would much prefer for his pocket mice.

Does a spacecraft orbiting the sun sound too fantastic?

It will probably be realized sooner than we might guess. It is commonplace that man's progress is exponential, that is, progress does not occur at a steady rate, but rather proceeds faster and faster as time advances. For example, in the field of biorhythms, more has been learned in the last twenty years than in the preceding two hundred. Similar rate increases can be cited from such widely different areas as steel production, world population, the non-agricultural labor force, world

[2] R. G. Lindberg, G. J. De Buono, and M. M. Anderson: "Animal Temperature Sensing for Orbital Studies on Circadian Rhythms," *Journal of Spacecraft and Rockets*, Vol. II (November-December 1965), pp. 986-8.

energy consumption, and technical manpower.[3] It is therefore none too soon to be thinking and planning for experiments on how plants, animals, and man will react in deep space.

How did Robert Lindberg become interested in this field in the first place?

Like so many other "clock-watchers," he came into bio-rhythms from another field, and because his observations in that field kept calling his attention to the effects or rhythms. After receiving his Ph.D. from UCLA in 1952, and working in Hawaii under a Pauley fellowship, he joined the Atomic Energy Commission as a research biologist. Stationed at a desert test site, he studied the effects of radioactive fallout on local desert mammals. One of these was the kangaroo rat, another a common desert mouse, and the third, and most interesting, the little pocket mouse, *Perognathus longimembris.*

Lindberg's routine procedure was to catch the animals as they scurried over the desert, and send them to laboratories at UCLA for autopsy determinations of the effects of fallout on the large chromosomes. And since the bioassays required autopsy, he had no reason to keep the animals alive after he had caught them. It was important, however, that they arrive at the laboratory with little change from their natural condition. To minimize such changes as might be brought on by the desert heat, Lindberg would thrust his captives into plastic bags and immediately place them in a chest containing dry ice. Presumably the animals were quickly frozen to death, or if not, then asphyxiated in the carbon dioxide atmosphere.

The animals so treated arrived in well-preserved condition at UCLA, and after they had had time to thaw out, technicians carried out the bioassays in routine fashion.

Then one day Lindberg brought his dry ice chest into the laboratory himself, dumped the frozen plastic bags onto a bench with a rock-like clatter, and sat down to chat with members of the staff. It was a long chat, for it continued until the bags were fairly well warmed up. Suddenly talk stopped

[3] Harrison Brown, James Bonner, and John Weir: *The Next Hundred Years* (New York: The Viking Press, Inc., 1957), *Passim.*

dead. Lindberg stared at the benchtop in utter disbelief. Several of the plastic bags were jumping up and down!

Lindberg snatched one up and peered inside. It was a little pocket mouse, and it was as lively and active as ever. Later it became fully apparent what an unusual creature the little pocket mouse really is. Although it is a warm-blooded animal, it will go into a state of torpor almost instantly on chilling, and once in that state it can be frozen stiff and still survive. When it is thawed out, its temperature rises rapidly, and it leaves the state of torpor and becomes active almost immediately.

What a wonderful animal for experimental purposes! Ideas churned through Lindberg's mind. Part of his graduate work had been done at Scripps Institution of Oceanography, and there he had developed a beginning interest in biological rhythms. Now he had right at hand an animal whose activity could be suspended—perhaps for long periods—by simple chilling. He could certainly expect that it, like other animals, would have a daily cycle of activity. Perhaps its body temperature might also display a conspicuous daily rhythm. In any case, it offered exciting promise as a new tool for studying such rhythms.

Meanwhile—on October 4, 1957—the Russians had launched Sputnik I into orbit around the earth, and by 1959 the United States had undergone an upheaval that ranged from revolutionizing methods of teaching grade school science and mathematics to feverish preparations for launching Discoverer 13 and recovering the first capsule from orbit.

Lindberg's imagination made the connection between the probability that man would soon place himself aboard one of his space vehicles, and an experimental animal that could survive low temperatures and the near-absence of oxygen. Here was a unique opportunity to contribute to space research, and in 1959 he joined the staff of the Northrop Corporation. He has continued there since, becoming head of the Bioastronautics Laboratory, and devoting much of his time and energy to developing the pocket mouse as a highly refined and sophisticated experimental tool.

The problems of handling the pocket mouse in space are negligible compared to those of any other animal known. *Perognathus* weighs about a third of an ounce—a fraction of the weight of a first class letter. It drinks no water at all and eats nothing but dry seeds. In consequence it excretes less than a drop of urine a day, and the feces are minuscule, dry, and odorless. Its oxygen requirement is so low that it can be soldered up in a tin can for long periods without harm. By chilling, it can be put into the dormant state for long periods, and it need not be aroused until it is needed for experiments. When needed, an increase in the surrounding temperature will arouse it immediately, and it is ready to go to work.

Its daily rhythms are clear and easily measurable. There are two of these: a rhythm of activity, and a rhythm of body temperature change. Of these, the one that gives the most satisfactory results is the rhythm of body temperature. The pocket mouse is what is called a facultative homeotherm, which means that in the active state it is a warm-blooded animal; its body temperature is only a little lower than man's. In torpor, its temperature drops to that of the surroundings. When active, its temperature ranges between about 93° and 97°F., and when it is in torpor its temperature drops to about 70° if it happens to be in an ordinary room. Or it may drop much lower, as we have seen, without harm to the mouse.

Lindberg has found that the time that a pocket mouse enters torpor spontaneously is a well-defined marker of its circadian rhythm, and that this is readily observed in the drop in body temperature. However, he has found that the marker is subject to aberrations from two sources. Sometimes an individual mouse will skip a day without entering torpor at all, but Lindberg has developed an autocorrelation method by which he can correct for that variable. Second, the time that a mouse enters torpor may be undesirably irregular; Lindberg overcame that with the finding that the time of arousal from torpor is far more precise. He now uses that as the marker for his rhythm studies.[4]

[4] R. G. Lindberg, J. J. Gambrino, and P. Hayden: "Circadian Periodi-

Lindberg's next step was to find a convenient, reliable method for recording the body temperature—one that could be used for earth-bound experiments, and that would also yield information that could be stored in spacecraft and transmitted back to earth on command. For this, the rapidly developing techniques of telemetry were ideal. Already being perfected were tiny radio transmitters that could be implanted in living animals, and used to study their behavior in their natural unrestrained conditions. Observations were underway on the free movements of Galapagos turtles, the heartbeats of birds in flight, and even the pressure variations inside the human eye for research on glaucoma.

Lindberg developed similar telemeters for recording the temperature variations and the activity periods of his pocket mice. The radio transmitter was less than a tenth the size of a penny, and it could be implanted in the abdomen of a mouse without bothering him in the least. Powered by a tiny mercury cell, the transmitter was sensitive to a tenth of a degree of temperature, and had a service life of two hundred and fifty days. For the proposed satellite experiments, Lindberg envisioned a receiver-recorder system that would take temperature readings every five minutes, and store it for transmittal back to earth once during each orbit.

He had now progressed to the point where he had a reliable experimental animal that was exceptionally well-suited to his purpose; and to go with it he had instrumentation for recording its rhythms of behavior under any experimental conditions that might be encountered.

Was he then ready to proceed with a full-blown independent research program? Should he now lay the groundwork on which future space experiments might rest? He decided that that would not be the prudent course.

He had learned a great deal about biorhythms, but he also knew that the planning and interpretation of experiments in

city of Resistance to Ionizing Radiation in the Pocket Mouse," *Proceedings of the International Biochronometry Symposium*, Friday Harbor, Washington, September 4–6, 1969. In press.

this field can be treacherous. He felt that to make the surest progress he should seek the advice and collaboration of a long-experienced authority in the field, and he turned to Colin S. Pittendrigh. Would Pittendrigh be interested in a cooperative arrangement? He would indeed. The possibilities sounded fascinating.

Although he was then at Princeton, Pittendrigh would soon be visiting Stanford, and he told Lindberg to meet him and bring some of his equipment with him for demonstration.

Carrying nothing but a slender briefcase, Lindberg walked into a restaurant near the Stanford campus on the appointed day and looked around for the vigorous man with the deeply tanned face and the shock of wavy snow-white hair. Pittendrigh greeted him with his usual outgoing enthusiasm, glanced briefly toward the bar, and then shook his head. The drinks were too expensive there, he pointed out, and suggested that they go to his place.

Settled in Pittendrigh's quarters, drinks at hand, the professor eyed Lindberg quizzically, reviewing quickly in his own mind what he had learned of Lindberg before agreeing to the cooperative arrangement. By all reports here was a man—six years younger than himself—who had already made outstanding contributions to biorhythm research, and whose career still lay mostly in the future. Here was a man with an easy, humorous manner that served only to highlight the depth of his insight and the clarity of his thought—a man of slighter build than his own muscularity, and with a relaxed ease of movement and a quick, infectious smile. Pleased, Pittendrigh flashed a question. Where was the mouse?

Right there in the briefcase, Lindberg told him. In a little soap dish. He took a tiny radio from his pocket and handed it to Pittendrigh. The professor listened to the radio for a moment, and let his glasses fall to the length of their rawhide thong. The mouse was telemetered. Very clearly Pittendrigh could hear him moving around.

So began a close collaboration that turned out to be so fruitful that Lindberg once told me, "I sometimes think I should have that briefcase preserved in bronze."

During the days that followed, Lindberg and Pittendrigh worked out a design for a complete animal and instrument unit that could be placed aboard any space vehicle, together with the design of a complete experimental program for it to carry out.[5]

This animal package provides six thin-walled aluminum cylinders 12 to 18 inches long and an inch and a half in diameter to house six pocket mice. Air at normal temperature and pressure would flow through the cylinders, and each mouse would be provided with a 30-day supply of loose sunflower seeds. Coarse screens at the exhaust ends of the cylinders would prevent the escape of seeds under weightless conditions, but permit the escape of seed particle and feces into a debris trap. An asbestos-cellulose lining would absorb urine, and provide for grooming, and a foothold for the mice; it would also serve as shielding for the electronic instrumentation. The aluminum tubes would be mounted in two layers of three tubes each, surmounted by an instrument package. Lindberg estimates that the complete self-contained unit for a thirty-day mission, incorporating data handling and storage, together with the environmental control system, would weigh no more than 25 pounds, would occupy no more than one cubic foot, and would consume only 12 watts of power.

Realistically, what is the chance that such a unitized biological laboratory will ever be put into orbit?

Our best indication of what may lie in the future comes from the National Aeronautics and Space Administration. In the early 1960's NASA asked the American Institute of Biological Sciences to set up five regional study councils, dividing the United States into five regions, to prepare recommendations for a program of biological research in a manned earth-orbiting space station. The Institute submitted those recommendations to NASA in October, 1966. After consideration of that report, NASA asked the Institute to proceed further by consolidating the original regional recommendations, and then assigning priorities to them according to time and scien-

---

[5] Lindberg, De Buono, and Anderson: "Animal Temperature Sensing," pp. 986–8.

tific merit. NASA also asked the Institute to define the environmental requirements of a space vehicle suitable for carrying out the recommended experimental program.

The five regional councils reviewed their earlier proposals, and then the council chairmen and three additional academic advisors met as a panel to prepare the final report. Chairman of that overall panel was Robert Lindberg.

The summary recommendations of the panel were these:

A. Members of the Regional Councils have displayed increasing enthusiasm and support of the Manned Space Station for biological research and specifically endorse the concept of an interdisciplinary experiment program encompassing many species of organisms representing various stages of phylogenetic development. Their response to the various rigors of space must be measured at several levels of cellular organization.

B. The highest priority recommendation made in the reports of the Regional Councils and reaffirmed by the Regional Council Chairmen is for strong ground-based biological research in support of space research. Ground studies are likely to provide the best and the most economical base for the development of essential long term information and skills required for the exploration of space and the advancement of biology.

C. The concept of investigator teams to conduct biological research in space is essential and should be developed to the fullest practical extent by NASA.[6]

Specifically with respect to biological rhythm research, the panel recommended as follows:

Living organisms exhibit rhythms in many of their functions. Manned space flights afford the unique opportunity to study the basic aspects of the role of sensory detachment on biorhythms. There are several classifications of biorhythms, such as diurnal, lunar, and annual rhythms. Biorhythms are influ-

[6] Robert G. Lindberg, et al.: *Bioscience Research During Earth-Orbiting Mission: Manned Orbital Space Station,* Final Progress Report to National Aeronautics and Space Administration, NASr–132. December 15, 1967 (Washington, D.C.: American Institute of Biological Sciences; 1967), p. xvi.

enced by both intrinsic and extrinsic factors (which may, in some instances, be artificially entrained) and may be best elucidated by comparative studies of representative organisms in a phylogenetic series.

Because many of these rhythms are linked to the terrestrial environment, it would be of immense interest and importance to study rhythms in the near-space environment. Such studies might provide basic information on mechanisms of biorhythms, as well as on the effects of sensory detachment. The latter aspect, of course, impinges directly on potential discomforts and dangers to space voyagers.

Despite the fundamental nature of the "biological rhythm" phenomenon, this category of research did not receive a high priority because of the divided opinion within the Councils as to whether or not experiments in Earth-orbit could be definitive. It was agreed, however, that rhythms are likely to affect the results of most biological experiments in the satellite and must be taken into consideration in the design of such experiments and in the analysis of the data which results from them.[7]

After it received that report, NASA set up a special task group of selected authorities that met in Carmel, California, in the summer of 1969. NASA also solicited informal opinions of the world experts who attended the International Biochronometry Symposium that was held at Friday Harbor, Washington, in September, 1969. This symposium was sponsored by the National Academy of Sciences-National Research Council, with NASA support.

NASA has not yet made public any decisions that it may have taken, but beyond any decisions of its own lies the whole arena of politics that involves competition for government funds. How that will turn out, only the future can tell.

In the long view, biological laboratories in deep space seem just as certain as today's realized fact that man has set foot on the moon.[8]

---

[7] Robert G. Lindberg, et al.: *Bioscience Research During Earth-Orbiting Mission*, p. x.

[8] According to Dr. Hubertus Strughold, Professor of Space Medicine and Chief Scientist, Aerospace Medical Division (AFSC), Brooks Air

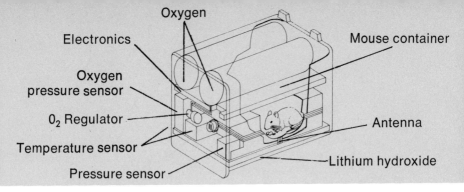

The Lindberg-Pittendrigh biospace package is a modification of the one described in the text; it holds one mouse, and measures about 5×8×9 inches. Because it can go so long with little food, water, or even air, the desert pocket mouse is an ideal animal for research on biological rhythms.

To be sure, delays may come at the hands of those who feel that less money should be spent for space research, and the money so saved used for the improvement of ecological conditions on earth. This, from the perspective of the history of science, seems short-sighted. As leading scientists have agreed in their testimony before a congressional committee: "We cannot afford an arrogance that may cause us to lose the possibility of major discovery." The immediate benefits of pioneering scientific research can only be surmised—they cannot be forecast.

Dramatic improvements in man's ability to manage the earth's ecology may well originate from new knowledge of how living organisms react to other environments—perhaps, even, the rhythms of the little pocket mousetronaut.

Force Base, Texas, "Mars will probably be the only post-lunar astronautical target for this century, unless new evidence is found by modern astronomy—such as rocket-and-balloon astronomy—that some of the other planets—Venus, in particular, are accessible to manned landing missions. Be that as it may, the physiological clock, manifested in circadian rhythmicity, will play an important role for health and performance in man's conquest of space. Furthermore, its study under non-terrestrial cosmic conditions may shed some light upon the true nature of this geobiological, or cosmobiological, phenomenon, which is still far from being completely understood." From "The Physiological Clock in Aeronautics and Astronautics," a paper presented before the Conference on Civilian and Military Uses of Aerospace sponsored by the New York Academy of Sciences, New York, N.Y., January 11–14, 1965.

*"After all, the dangers
that face the world,
every one of them,
can be traced back to science.
The salvations that
may preserve the world,
every one of them, will be
traced back to science."*

ISAAC ASIMOV,
1970[1]

20. Stress and the Space Age

More than a hundred years ago, in his essay "Economy," Thoreau said that "he never assisted the sun materially in his rising." Today a man may *cause* the sun to rise. All he needs to do is board a fast jet headed east across many time zones, and the sun will rise in the *west*.

The explanation is simple. In the latitude of a flight from New York to Paris, daylight proceeds around the globe at a speed of about 800 miles an hour—a shade over the speed of sound. Therefore at somewhat more than the speed of sound, the eastbound plane will outrun the speed of the earth's surface, and if the plane's departure is properly timed, the sun will rise in the west during the flight.

It is hard to realize that we have come so far so fast. As of the spring of 1970, the United States had carried out

[1] Isaac Asimov: "Science Fiction, an Aid to Science, Foresees the Future," *Smithsonian*, Vol. I (May 1970), pp. 41–7.

twenty manned orbital space flights, the first of these having taken place only eight short years before. It was as recently as February 20, 1962, that John Glenn, Jr. flew "Friendship 7" three times around the globe and landed safely in the sea. To achieve orbital flight, he had to reach the "first cosmic velocity" of 8 kilometers per second, or some 18,000 miles per hour.

Compare this with the speeds of commercial aircraft just after World War II. In 1946, four-engined transport planes were flying at 370 miles an hour, and there were still some doubts as to whether the "sound barrier" could be broken —that is, whether flight speeds greater than the speed of sound (760 mph) could ever be achieved. Twenty years later, Hubertus Strughold was seriously discussing the advantages that diplomats would have in making round trip flights in a single day from Washington, D.C. to the capital cities of Europe and return. Such flights would require speeds up to five times the speed of sound, or some 3,800 miles an hour (Mach 5 in astronautical jargon).[2]

How might flight at such speeds affect man?

The answer involves two different but closely related questions. The first question concerns what we may call "earthbound" flights at speeds from about Mach 1 to Mach 5— that is, flights intended to get from one place to another on the surface of the earth. The second question concerns flights intended to put a vehicle into orbit, or into deep space.

The first indications of man's difficulties with fast earthbound flights came as a result of painful practical experience.

In 1960 S.A.S., the Scandinavian airline, began passenger service between New York and Copenhagen using Douglas DC-8 jets. The same year they pioneered the polar route from Copenhagen via Anchorage to Tokyo. But it was some time before news began to filter out that passengers were

[2] Hubertus Strughold: "The Physiological Clock in Aeronautics and Astronautics," (Paper presented before the Conference on Civilian and Military Uses of Aerospace sponsored by the New York Academy of Sciences, New York, N.Y., January 11–14, 1965), p. 10.

suffering something more deep-seated than the aftermath of farewell parties.

An early article that shed light on what was happening was written by Samuel W. Bryant for the November, 1963, issue of *Fortune*. Reporting on a flight over the polar route Bryant told of one man's experience:

"The airplane door on the Tokyo-Anchorage-Copenhagen flight bangs shut, the passengers fasten their seat belts, and the captain shoves the thrust levers forward. When the big jet leaves the runway, it is 9:30 P.M., Tuesday, Tokyo time. Three hours later the flight meets the rising sun and three and a half hours after that touches down at Anchorage, where it is 9:00 A.M. and—thanks to crossing the international date line—still Tuesday. At 11:00 A.M. begins the eight-and-a-half-hour flight over the polar wastes to Europe, in bright summer sunlight. When the plane lands at Copenhagen it is 6:25 A.M., Wednesday. The elapsed time since leaving Tokyo has been almost seventeen hours—nearly all of it in daylight—and as any customs agent can plainly see, when the passengers debark they are confused and dead tired.

"One recent victim of this 'over the top' speed, Kazuma Nakayama, of Nomura Securities Co., Ltd., went on to catch a 7:45 A.M. plane for London, where it was 9:25 A.M. on arrival. 'When I got to London,' says Nakayama, 'I was tired and my body was insisting that it was time to go to bed. But my sense told me it was midmorning and time to get to work.' "[3]

A few months after the *Fortune* article was published, Lowell Thomas was a guest on an Arthur Godfrey anniversary radio program, and Thomas alluded briefly to an illness from which he had recently recovered, and which—according to his doctors—had been brought on by his having "crossed too many meridians too fast." He has been kind enough to provide me with the firsthand story of what happened.

"Enclosed are some lines," says his letter, "that I wrote

[3] Samuel W. Bryant: "What Jet Travel Does to Your Metabolic Clock," *Fortune*, Vol. LXVIII (November 1963), pp. 160–8, 183–6.

at the time that I was knocked out, just in case they might be of some use to you. Since the Detroit incident, when I was hospitalized, I have been around the world again, in fact twice, and a second time to the South Pole. On my most recent journey I paced myself, and did my utmost to get enough sleep; and, I watched the problem of eating and drinking."

Here is his story:

"Two weeks ago, the day I flew in from the South Pole, I was involved in a banquet where all the speakers had a rough time because of the acoustics and the hullabaloo going on at the bar in an adjoining room. All this for four hours, with me as the climax speaker! Incidentally, I had made two other speeches earlier that day. At any rate, when the late banquet was over I felt groggy and decided to have a check-up. Maybe it was my lucky night. Dr. Herman Alvarez, who is a wizard in this field, tells me that if I'll follow instructions I may be around for another twenty or thirty years, able to go on a dozen more expeditions, and even get to this year's winter Olympics, at Innsbruck.

"The doctors seem to think I've been racing my motor more than usual this year. They say three jaunts around the world, two expeditions to the Antarctic, one to the mountains of New Guinea, another to Central Africa, a journey to the Himalayas to see the Dalai Lama in exile, a trip to Central Asia—they think these journeys, added to my usual ski trips to New England, the Rockies, and New Zealand, along with radio, TV, and film chores, plus a book, is—well, they say I should slow down. So I guess I'll have to give up golf!

"As General Jimmy Doolittle wired me at the time, 'don't give up the jet speed, Lowell. Just turn off your after burner.' "[4]

Even in 1965, however, it was growing more and more apparent that something more than just flight speed—indeed something even more than Lowell Thomas's hectic pace—

[4] Lowell Thomas to Ritchie R. Ward, personal communication, March 26, 1965.

was causing the upsets that jet passengers were feeling.

An important clue had come from the experiences of jet pilots. They agreed that flights from west to east were more upsetting, while those from east to west were only slightly less so. In sharp contrast, flights from north to south or back—at the same speeds and over the same distances—disturbed them not at all. As Bryant pointed out in his *Fortune* article, flights crossing many meridians threw man's biological clock out of phase with his environment, while flights up and down meridians kept within a single time zone, and the clock remained in phase with the environment.

In my own view, it seemed likely that throwing the clock out of phase must cause a stress reaction of some sort, but the cause of the stress, and more important, any possible

A passenger on a fast jet flight from Copenhagen to New York arrives at his destination tired and upset, because his biological clock has been thrown five hours out of phase with his environment. But after flying at the same speed and about the same distance from Santiago, Chile, to New York, he arrives rested and refreshed; he has remained in the same time zone throughout, and his clock is undisturbed.

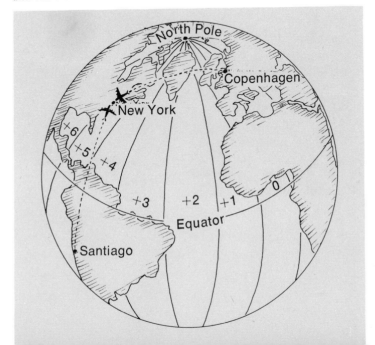

remedy, remained vague. There was no hint of any physiological basis for the disturbance other than the superficial observation that man was timed to his environment, and if he changed that timing, he could expect trouble.

Knowing that Dr. Hans Selye, of the University of Montreal, was a world authority on stress in relation to disease, I consulted two of Selye's own works: *The Stress of Life* (New York: McGraw-Hill, Inc.; 1956) and *From Dream to Discovery* (New York: McGraw-Hill Inc.; 1964), but found nothing on stress in relation to jet flight. Accordingly, I put the question to Dr. Selye direct, and received the following answer:

In reply to your inquiry of March 20th let me say that I have done no work on the alleviation of stresses caused by rapid changes through many time zones, and, to the best of my knowledge, no one else has published on this topic either. Now that you mention it, I wonder why this problem has not received more attention in view of the enormous literature concerning stress in relation to other problems of aviation and space medicine.

Since I have too often been exposed to this kind of stress myself, I hope that you can interest someone to do research along these lines.

> *Very sincerely yours,*
> */s/*
> *Hans Selye*[5]

Fortunately we soon learned that research along these lines had already been done. It was not available in published form at the time that Dr. Selye wrote, but only a few weeks earlier it had been reported by Dr. Hubertus Strughold, in the paper cited and it offered jet travelers some practical solutions to their problems.[6]

Dr. Strughold set the tone of his discussion by pointing out that one of the most impressive of all life's cyclical processes is the alternation of sleep and wakefulness, and that that cycle is associated with the physical cycle of day and

[5] Hans Selye to Ritchie R. Ward, personal communication, March 23, 1965.
[6] Hubertus Strughold: "The Physiological Clock," *passim*.

night caused by the rotation of the earth. The sleep-wake cycle, he said, has become so firmly established as a vital feature of living organisms under the earth's influence that, in man's case, to break it completely leads to serious neurotic disturbances.

This would place man's clock upsets somewhere between those of Frank Brown's oysters and Janet Harker's cockroaches. Recall from Chapter 15 that after Brown had transported his oysters from New Haven, Connecticut, to Evanston, Illinois, the oysters followed the New Haven tide table upon arrival, but that during the course of two weeks, they gradually adjusted themselves to the tides that would occur at Evanston if it were a coastal city. There was no evidence that the oysters were discommoded in any way.

Janet Harker's cockroaches, on the other hand (Chapter 13), had no way of making any adjustment, because Harker's cell transplants had left them with two clocks, one running completely out of phase with the other. Under stress of this severity, the roaches quickly developed cancer and died.

Strughold illustrated his point about man's clock disturbances by recalling that Napoleon I, Emperor of France, and Frederick the Great, King of Prussia, both had the idea that sleep was just a bad habit, and both set out to break it. Neither lasted for more than two nights. They were trying to escape a behavior pattern that had evolved under the influence of the earth's cycle of day and night over a span reaching back before the dawn of history, and in their day, they might as well have tried to defy the law of gravitation.

Recognizing the inescapable nature of the forces involved, Strughold made no attempt to combat them, but sought out practical measures that could afford jet passengers some relief from their upset clocks.

He arrived at three specific recommendations:

First, if the traveler needs to be fully alert at his destination, a simple solution is for him to arrive several days in advance, and take plenty of rest during his period of adjustment. This would be of the greatest usefulness before important military, diplomatic, or business conferences. Unfortunately, however, men involved in such conferences would be those least likely to be able to spare the time for such adjustment.

Second, the physiological clock could be preset, so that upon departure it would be in phase with the time zone of the destination. This would require the traveler to spend several days gradually shifting his sleep-wake cycle, and of course during such an adjustment period, he would get out of phase with his own social regime.

Third, Dr. Strughold suggested mild medication, under the supervision of a physician sufficiently aware of the basic problem to be able to prescribe the right medication, at the right time, and in the right dosage.

In his capacity as professor of space medicine in the U.S. Air Force, Dr. Strughold was naturally most concerned about the problems of government personnel; and they have the advantage of having ready access to geographical and medical flight information.

What then of the more usual business or vacation traveler? How may he learn how much upset could result from a flight he plans, and what measures he should take to minimize his discomfort?

His first step would be to find out how many time zones separate his point of departure from his destination. If the time difference is no more than the three hours between San Francisco and New York, then he may expect some discomfort, but it should not last for more than a day. If it is the nine-hour difference between San Francisco and Copenhagen, the upset will be substantial, and will last for several days. If the difference is as great as that experienced by Kazuma Nakayama, then disorientation as severe as his can be expected.

Time-zone differences between major cities can be approximated very easily if one is at an airport that has a bank of clocks set at various geographical times. It is a simple matter to relate the differences shown by such clocks to one's own plans.

Or, if the traveler is looking over travel folders at home, he will undoubtedly have a world map at hand, from which he can approximate the longitude of any city. All he then needs to know is that the geographic time difference from point to point is one time zone for every fifteen meridians, or a total of twenty-four time zones in all. He will see, for example, that San Francisco is at about 122° West, and New York is at about 74° West, and that the difference is 46°. This divided by fifteen gives three times zones, which of course is correct. Again, Copenhagen is at about 12° East, and to find the difference between San Francisco and Copenhagen, the 12° East must be *added* to the 122° West of San Francisco. The 134° sum, again divided by fifteen, gives nine time zones.

Of course actual time zones do not follow the meridians exactly, but the deviations are not large enough to make any practical difference for this purpose.

The business or vacation traveler can thus know in advance how seriously his own clock may be disturbed by a planned flight, and he can adapt Dr. Strughold's recommendations to his own situation. In advance of especially long flights, he might do well to consult his own physician for advice on diet, sleep before, during, or after flight, and mild medication.

Of course Dr. Strughold recognized that the measures that he suggested were simply palliatives, designed to give relief for temporary upsets. For the future, he suggests that the whole problem (at least for those whose business in a distant place would require only a few hours) may well disappear with the advance of technology. When a man can go halfway around the world and back in a single day, he can be back in bed in the same time zone in which he arose, and he should not then experience any discomfort, much less any disruption.

But that is for earthbound fliers. The problems of orbiting astronauts, and especially of those who will penetrate into deep space, are of an entirely different order. The only hope for their eventual solution lies in long-range fundamental research.

To follow today's search for new knowledge in that field, we shall visit a soundproof underground operation room left over from World War II.

21. Aschoff: Man's Own Clock

The scientists we have met so far, and the discoveries they
made, have had this in common: a question or problem was
set forth, an experimental method for solving it was devised,
and an answer was found.

Charles Darwin asked why leaves sleep, he pinned out
leaves on cold nights so they could not fold, and from his
observations of their reactions he concluded that leaves
sleep to protect themselves from the chill of the night sky.
Janet Harker asked where in an animal its living clock might
be found, she carried out surgical transplants on cockroaches,
and she found that the cockroach clock lies in four cells in
the roach's secondary brain. Karl Hamner asked if the forces
associated with the earth's rotation were responsible for diurnal
rhythms, he took plants and animals to the South Pole, spun

them opposite to the earth's rotation there, and from his observation that the rhythms persisted, he concluded that they were not caused by forces associated with the earth's rotation.

In these cases it was possible to formulate a specific question, devise an experimental method, and obtain an answer to the question.

If we wish to proceed, then, from lower organisms to research on man, what questions would we ask about man's own clock?

A moment's reflection will show that merely to list them would be a formidable task, much less to devise experimental methods to test them, and means of interpreting the results.

The magnitude of the problem was recognized more than thirty years ago by John H. Welsh, who wrote: "So many factors have been found to affect the 'diurnal curves' of man that the conclusions are most confusing. Man is a poor experimental animal, and at present, it would seem that a more complete understanding of the internal processes responsible for diurnal rhythms would come first from study of the lower animals."[1]

A modern view of the same difficulty has been given by Jürgen Aschoff, whose work in the field has been outstanding. He was a member of the organizing committee for the Cold Spring Harbor Symposium on Biological Clocks in 1960, and gave the keynote paper at the session on the general properties of circadian rhythms. He edited the Proceedings of the Feldafing Summer School on Circadian Clocks that was sponsored by the Scientific Affairs Division of NATO in 1964 and contributed two important papers there.[2] He gave the opening paper—on interdependence of parameters of circadian rhythms of birds and man—at the Friday Harbor

[1] John H. Welsh: "Diurnal Rhythms," *Quarterly Review of Biology*, Vol., XIII (1938), pp. 123–39.
[2] Jürgen Aschoff, ed.: "Circadian Clocks," *Proceedings of the Feldafing Summer School, 7–18 September 1964* (Amsterdam: North Holland; 1965).

In his laboratories at Stanford University, Colin S. Pittendrigh (left) explains to Jürgen Aschoff the operation of the highly sophisticated equipment that he has designed for studying the rhythms of fruit flies. (Photo by the author).

authority on sleep, and Bruce Richardson, of the University of Chicago, shut themselves up in a chamber of Mammoth Cave, Kentucky, in 1938. The object of that experiment was to see if either of them could adjust his clock to a cycle ther than 24 hours. They spent thirty-two days in the cave an attempt to adapt to a 28-hour cycle, divided into 19 hours activity and 9 hours of sleep. By the end of the first week hardson, who was then twenty-three, had successfully his clock to the new schedule, but Kleitman, who was years older, could not make the adjustment.[4]

his was an interesting observation in its day, but in ison with the kind of modern facility that we have d, the Mammoth Cave would be like the Garner-Allard g house" (page 116) compared to the Canberra (pages 128-9).

Goudsmit and Robert Claiborne, eds.: *Time* (New York: oks; 1966), p. 52.

Biochronometry Symposium that was sponsored in 1969 by the National Academy of Sciences-National Research Council with support from the National Aeronautics and Space Administration. If his colleagues were asked for a one-word description of him, they would probably say "imaginative."

Holding an M.D. degree from the University of Freiburg, Aschoff is uniquely qualified for his research on human rhythms. His M.D. thesis was on blood alcohol and habituation, an early indication of his interest in physiology. He practiced briefly at the clinic for internal medicine at Frankfurt am Main, and then joined the physiological institute at the University of Göttingen. During the academic years 1948 and 1949 he gave the principal lectures in physiology at the medical school of the University of Würzburg, and was named professor of physiology. He now maintains his academic association as a member of the medical faculty at the University of Munich.

"It is well known even to the layman," Aschoff wrote, "that each day the body temperature reaches a highest value toward the evening and a low point early in the morning. Since this phenomenon was described by Gierse in 1842, numerous clinical and physiological studies have shown that there is apparently no organ and no function of the body which does not exhibit a similar daily rhythmicity. Whether we measure, hour by hour, the number of dividing cells in any tissue, the volume of urine excreted, the reaction to a drug, or the accuracy and speed with which arithmetical problems are solved, we usually find that there is a maximum value at one time of day and a minimum value at another."[3]

On Welsh's first point, why *should* man be a poor experimental animal?

Without embarking on a philosophical discussion on how

[3] Jürgen Aschoff: "Circadian Rhythms in Man," *Science*, Vol. CXLVIII 148 (June 11, 1965). pp. 1427–32. Elsewhere Aschoff has said that in man, more than one hundred functions and structural elements could be named which oscillate between maximal and minimal values once each day.

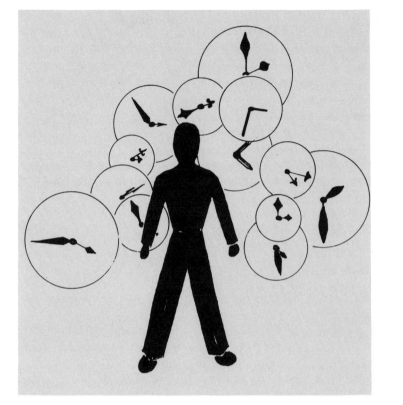

In addition to his obvious sleep-wake cycle, man exhibits daily periodicities in the rates at which his cells divide, his physiological functions, his reactions to drugs, and his speed in solving problems. As many as one hundred functions show daily cycles.

man differs from the lower organisms, we may note for our present purposes that man enjoys beating the rap; if the consequences do not seem too threatening, he will bend his IBM card. The timing of his behavior in an experimental situation is far less consistent than that of a deer mouse. Thus, one subject in a carefully controlled experiment designed to measure his free-running activity rhythm deliberately set out to see if he could lengthen his days, and succeeded in doing so.

And, of course, Welsh's point that many factors affect the

rhythms of man is still valid. A great many of the experiments that he would have liked to see on lower animals have now been done, and a more complete understanding of the internal processes involved has been reached. Still, understanding man's clock remains the knottiest problem facing rhythm researchers today.

That may be one of the reasons why not more than a handful of them have tackled it. But a much more important reason is that it is difficult to obtain subjects for experiments, and it is time-consuming and expensive to obtain enough experimental data to have statistical reliability.

One can breed fruit flies by the hundreds of thousands at small expense, and obtain data having good statistical confidence limits. One can germinate bean seedlings without limit. One can feed scores of little pocket mice for week on a handful of sunflower seeds.

In contrast, consider the problems involved in car out research on the rhythms of human subjects. Bear i that to obtain results comparable to modern experime lower organisms, it would be necessary to provid for maintaining precisely controlled conditions of perature, humidity, and so on. All external cue any notion of true time could be derived wo excluded, in particular, noise of any sort. And most important time-giver, it would be e precisely controlled light levels. It wo sary to provide for all of a subject's ph out giving any cue as to the time, him to occupy his time during isola

To follow the rhythms unde necessary to provide instrume of sleep and wakefulness, bod clinical facilities for determir all this for one individual subject could not influenc

Now it is true tha have long since isola example, the human slee

In short, such a facility would be difficult to provide, expensive to maintain and operate, and slow to yield results. It is therefore no surprise to find that there is only one of its kind. It is located at the Max-Planck-Institute for Physiology of Behavior at Erling-Andechs, Germany, some ten miles south of Munich. There the isolation facility is known as the *Tier Bunker,* from the German *Bunker,* which translates approximately to the English word of the same spelling, and *Tier,* "animal." It is under the direction of Jürgen Aschoff, who first had the concept for it, and who is director of the Institute's division that bears his own name: "Abteilung Aschoff."

Aschoff's direct interest in human rhythms resulted from a series of studies that he made on caged chaffinches, but his interest in animal rhythms began long before that.[5] More than twenty years ago he determined the activity rhythms of mice in constant light and constant darkness, and his extensive studies of the effects of light on these and other animals resulted in a hypothesis that has now become widely known as Aschoff's Rule. This says that as light intensity is increased the activity period of a day-active animal will decrease, and that of a nocturnal animal will increase. The rule has been widely confirmed by independent workers for many different species, although occasional exceptions have been noted.

In his studies with finches, Aschoff found the rule to be well supported. He used the finch's characteristic activity of hopping from one perch to another (recorded automatically by means of printout counters) as the "hands of the clock," and kept the birds under constant conditions except for the experimental variations of the light regime. In a typical experiment, he exposed the finches to the sequence of light conditions listed below, with results as noted:

I. Entrainment to alternating light and dark conditions with 12 hours of light and 12 hours of darkness (LD 12:12). As soon as the lights were turned on, the birds began hopping from perch to perch, and continued throughout the

[5] The chaffinch is the common finch, often kept as a pet in Europe.

12-hour light period. As soon as the lights were turned off, they stopped.

II. Free-running in continuous *dim* light (0.4 lux) for 27 days. On the first 2 days the birds began their activity at the same time as before. But after that they began later and later, so that when they were running free, their clocks *lost* about an hour a day.

III. Reentrainment in alternating light and darkness (LD 12:12) for 15 days. The birds immediately ceased losing time and took up the same accurate 12-hour rhythm as in I.

IV. Free-running in continuous *bright* light (120 lux). The birds now began perch-hopping earlier each day, their clocks *gaining* nearly 2 hours a day.

This sequence confirmed the part of the Aschoff Rule that states that a day-active animal has a free-running period longer than 24 hours in dim light, and shorter than 24 hours in bright light.

Much more basic, it provided clear and convincing evidence that the chaffinch has a true *internal* clock—indeed a clock that can be reset by varying the light regime; it can be made to run on an exact 24-hour cycle, or slower or faster as desired.

With such evidence at hand, Aschoff became intensely curious to learn how man's own clock would behave under controlled laboratory conditions, and whether it also could be shown to be endogenous.

He was, of course, familiar with the popularly reported observations about man's unaided time sense. The story, for example, of the gypsy boy who knew the time of day almost to the minute, even though he had never owned a watch. And that a man can awake at a chosen hour in the morning if he fixes the idea in his mind the night before. And the far better supported observation that shift workers have rising and falling rhythms of efficiency with the time of lowest efficiency occurring at about 3 A.M.,when the highest frequency of errors and industrial accidents occurs.

What lay behind all this?

As we noted earlier, the problem is far-reaching and

extraordinarily complicated. Would there be any hope of obtaining meaningful answers short of a years-long program of intensive fundamental research? Aschoff thought not. And he started laying plans for a program extending far into the future.

The first place for isolating human subjects that he was able to lay hands on was a soundproof operations room left over from World War II. This was a makeshift, but it did produce some preliminary pilot results that encouraged him to plan for a more sophisticated facility. Those plans came to fruition with the completion of the *Tier Bunker*.

In the bunker the subjects live in complete isolation for periods of up to three or four weeks. Adjoining a comfortable bed-sitting room there is a shower, and a small kitchen in which the occupant prepares his own meals. "We ask him," explains Aschoff, "to lead a 'regular' life, that is to have three meals in a normal sequence, not to nap after lunch, and to perform a few psychological tests. Otherwise, he is allowed to do what he wishes. Many subjects are students who cram for an examination. Listening to recorded music is their favored entertainment."

All measurements of the subjects' activities and physiological reactions are recorded outside the bunker. These include a continuous record of body temperature by means of a rectal probe, a record of activity pattern, the subject's best estimate of how long it takes for 20 seconds to pass, and for an hour to pass, and his movements when in bed. The subject also collects all of his urine at such times as he chooses.

To ensure perfect isolation, the entrance to the bunker is provided with double locked doors equipped with magnetic catches, so that only one door at a time can be opened. Between the two doors, a small enclosure contains an icebox for interim storage of the urine specimens. The occupant's only means of communication with the outside is by written notes that he leaves in the enclosure. Through it he also receives fresh food supplies, and "other necessities such as one bottle of Andechs beer per day."

The experimenters open the outside door at completely

random times, so the occupant has no clue from that activity as to what time it is. Usually the occupant turns his lights off when he goes to bed, and turns them on when he gets up, but the intensity of the light is regulated from the outside. Aschoff allows the subject to regulate the temperature in the bunker, for he has found that small variations in this have no appreciable effect on the results.

Under these carefully controlled conditions all subjects displayed very clear circadian rhythms. These included cycles of sleep and wakefulness, body temperature, urine excretion by volume, and by calcium and potassium output. All of these functions were not in perfect phase with each other, but on average, the free-running period was 25.0 hours—that is, the free-running clocks were slow by an hour a day.

Aschoff himself was the first subject. He is not the kind of man who would seem likely to adjust to complete isolation very happily, for he is extremely active and gregarious. A rather slight, slender man with graying hair, he has an active inquiring mind that demands continuous exchange of ideas with colleagues. He has a keen sense of humor, which he enjoys sharing with others, and which is especially amusing to American ears when he uses his British-flavored English. But, gregarious as he is, Aschoff wanted to be the first in the bunker, to find out at first hand what timeless life would be like.

After he came out, he gave the following account:

"After a great curiosity about 'true' time during the first two days of bunker life, I lost all interest in this matter and felt perfectly comfortable to live 'timeless.' From the knowledge of animal experiments I was convinced that I had a period shorter than 24 hours; when I was released on day 10, I was therefore highly surprised to be told that my last waking-up time was 3 P.M. In the 'mornings' I had difficulty in deciding whether I had slept long enough. On day 8, I got up after only 3 hours of sleep. Shortly after breakfast I wrote in my diary: 'Something must be wrong. I feel as if I am on dogwatch.' I went to bed again and started the day anew after three more hours of sleep. Judging from the curve of body temperature, my first start happened to coincide with the worst phase of the circadian period, that is, with the low point of temperature. I was mistaken by an effort of will and put to order by my physiological clock."[6]

A number of years later, Aschoff told me of a dream that he had during one of his first nights in the bunker. In addition to the dislocation of his normal routine, he was much worried about the success of the experiment. This wartime surgical operations room in which he had placed himself was only a makeshift to serve for a pilot experiment. Would the outcome support his request for a large sum of money to build a real bunker? Or would outside noises reach him in the makeshift room, give him a clue as to the time, and ruin the experiment?

Under this stress he fell into a fitful sleep, and dreamed that he was in that very room, sitting at a table writing notes. As he did so, he was continually distracted by street noises: the clang and grind of street cars (there are no street cars in Erling-Andechs), the honking and exhaust noises of cars, and all the other background noises of a large city (there is no large city that close to Erling-Andechs).

He awoke in a cold sweat and listened intently. Only dead silence. No external clue had reached him. The experiment should succeed. He lay back and slept soundly.

The second occupant of the bunker was Patricia De-Coursey, who by then had finished her work on the clock of the flying squirrel (Chapter 11). She had accompanied her husband during his postdoctoral fellowship at the University of Tübingen, and then for six months at the University of Munich, and her reminiscences give us further insight into Aschoff's character:

"I participated as Aschoff's second *Bunker Tier*—i.e., Aschoff himself was the first subject in his now-famous human rhythms research. I think it was very hard on someone of his temperament to stay in isolation even for a few days, and he was eager to find a placid, solitude-loving subject for a

[6] Jürgen Aschoff: "Circadian Rhythms in Man," p. 1429.

longer duration experiment. I retreated happily with record player, notes, and plans for a new paper to a '3-week undisturbed vacation' in the bunker.

"During the last six months of our Germany visit, while we lived near Munich, near Aschoff's lab, I worked daily at his Institute. The association with him, in terms of stimulating ideas and new insights has had a profound effect on all my later work."[7]

Aschoff himself told me that DeCoursey had had one bad experience in the bunker. Toward the middle of her three-week stay, the thermostat broke down, and the bunker became unbearably hot. And those on the outside did not notice what had gone wrong. "Although Pat started out calm and collected," Aschoff said, "she really got disturbed then, for she had no way of knowing what had happened. But she got a grip on herself, and stuck it out, and I certainly admired her for not losing her cool."

By 1965, Aschoff had completed experiments on 26 human subjects. He considered the results of these observations to be preliminary and tentative, and we shall therefore only summarize them briefly here:

First, 22 of the 26 subjects showed periods longer than 24 hours.

Second, the periods were lengthened in dim light and shortened in bright light. Thus, since man is a day-active animal, they followed Aschoff's Rule.

Third, one subject was able to manipulate his period. The light level to which he was exposed was very high (1,500 lux), and during his first ten days of free-running, his period was only 19 hours. Then he tried consciously to extend his activity periods, and was successful in increasing them to 25.6 hours. Aschoff noted that since conscious control is possi-

---

[7] Patricia DeCoursey to Ritchie R. Ward, personal communication, September 20, 1969. DeCoursey did not mention how her clock compared to those of other *Bunker Tiere*. Personally she remained a dear friend of Aschoff's, but as an experimental subject she became just another *Bunker Statistik*.

ble for one subject, other subjects may show variable results because some degree of unconscious control might occur.

Fourth, one subject showed marked desynchronization of his own functions. His sleep-wake cycle was a surprisingly long 32.6 hours, while his body temperature and urine excretion rhythms were only 24.7 hours. Note, however, that even though the curves of two rhythmic functions have different periods, there will always be times when the curves cross— that is, times when the functions are in perfect phase. This particular subject kept diary notes, and the times when all of his functions were in phase were just those times when he felt especially fit and well!

The observation that internal desynchronization to this degree may occur in an occasional subject, *in the course of exploratory fundamental research,* led Aschoff immediately to the practical problem of desynchronization of jet passengers upon arrival in a distant time zone.

Does basic research offer any solution to this practical problem?

Aschoff thought that it offered the germ of an idea.

"Manipulation of artificial *Zeitgebern* (time-givers)—for example, the insertion of relatively short times of light and dark after extremely long flights—might be a . . . possibility. [But] these problems [have] only reached a state of theoretical analysis and of preliminary experimentation with animals."[8]

Well and good. Aschoff's basic research had led him to a point at which he could suggest a possible way to shorten the time a jet passenger would need to adjust to a new time zone. He also had excellent facilities for testing the idea, and long experience in planning and carrying out experiments with human subjects. Did he then rush forward in pursuit of immediate results?

He did not. He returned to experiments with chaffinches.

---

[8] Jürgen Aschoff: "Circadian Rhythms in Man," p. 1432. Recall, in this connection, that other investigators, with completely different objectives, had been able to reset both plant and animal rhythms by means of brief light flashes.

In the summer of 1970 I asked him if he had since obtained further ideas on the desynchronization problem. He shrugged. "Not really. Not since the 1965 *Science* paper. You see, we don't know enough about what really happens to people to design meaningful experiments. And we won't be able to, until we understand more about the fundamental nature of the process."

But with chaffinches he saw that it would be possible to simulate—right in his own laboratory—the same time zone displacements that jet passengers experience. He need only place the birds under a normally alternating light-dark cycle for long enough to entrain them to that rhythm, and then shift the cycle by, say, six hours. The effect should be the same as though the subject had suddenly been transported through six time zones; shortening the cycle would simulate eastward flight, and lengthening it would simulate westward flight.

With finches, Aschoff found that the readjustment period after eastward "flight" was about three days, and after westward "flight," about six days.

With two human subjects in the bunker, Aschoff found that the rhythm of activity was restored rather rapidly after "flight" in either direction, but that the rhythm of body temperature took much longer to adjust. As with finches, these human subjects adjusted more slowly after westward "flight" than after eastward. This observation is the reverse of the experiences of jet pilots, who say that westward flights upset them less than eastward. However, Aschoff points out that resynchronization after westward flight is faster for those who have long natural periods, and since many human subjects tend to have rather long periods under constant conditions, a large percentage of the population may resynchronize more easily after flight in the westward direction.[9]

What then of Aschoff's fascinating suggestion that artificial exposure to short light-dark cycles might shorten a jet passenger's adjustment period?

[9] Jürgen Aschoff: "Desynchronization and Resynchronization of Human Rhythms," *Aerospace Medicine*, Vol. XL (1969), pp. 844–9.

Jürgen Aschoff (left) and Colin S. Pittendrigh amuse themselves with a doodle of a free-running rhythm before getting down to the serious business of writing a chapter on biological rhythms for the three-volume Foundations of Space Biology and Medicine, *to be published jointly by the U.S. National Aeronautics and Space Administration and the U.S.S.R. Academy of Sciences. To work with Pittendrigh on this project at Stanford University in the summer of 1970, Aschoff took leave from his laboratory in Erling-Andechs, Germany, where he is Director of the Max-Planck-Institute for Behavioral Physiology.*

As of this writing, he has not published any findings, and we shall have to wait and see. As de Mairan pointed out more than two hundred and fifty years ago: "The advance of true natural philosophy, which is experimental, can only be painfully slow."

Even further into the future lie the answers that basic research will provide for the synchronization problems of the astronauts who will plunge into deep space. As to one source of those answers, we may take it as suggestive that Jürgen Aschoff's work is being strongly supported by the U.S. National Aeronautics and Space Administration.

That period of twenty-four hours,
formed by the regular revolution of our earth,
in which all its inhabitants partake,
is particularly distinguished in the physical economy of man.
This regular period is apparent in all diseases;
and all the other small periods,
so wonderful in our physical history,
are by it in reality determined.
It is, as it were, the unit
of our natural chronology.
CHRISTOPHER WILLIAM HUFELAND,
1796

These timing devices merit great interest,
not only because of their remarkable nature,
but [also] because of the light that they might throw
on the functioning of the various organs of the body,
particularly the different parts of the brain,
and on normal and abnormal behavior
of the total organism.
CURT PAUL RICHTER,
1959

22. Beyond Tomorrow: The Conquest of Disease

It is now nearly half a century since Paul de Kruif wrote the inspiring *Microbe Hunters,* no doubt the most widely read book in print about man's conquest of disease.[1] In it, de Kruif told the story of the germ theory of disease, and how the microbe hunters from Louis Pasteur to Paul Ehrlich developed this budding theory, and used it to fight age-old scourges from rabies to syphilis.

How marvelous it would be if, today, we could tell a similar triumphant story about the conquest of those diseases that are periodic—this host of maladies of body and mind

[1] Paul de Kruif: *Microbe Hunters* (New York: Harcourt, Brace and Company; 1926).

that seem to be so clearly related to our periodic physiological functions, and to our variable mental and emotional states.

But the time to tell this story has not yet come.

Without questioning the brilliance of the advances that de Kruif's early microbe hunters made, we must concede—as we approach the closing quarter of the twentieth century—that the germ theory of disease is really quite simple. From today's perspective, we might liken it to Newton's theory of gravitation, another outstandingly useful but simple theory.

Today, to solve our infinitely more complex problems, we need a new theory of how and why our clocks behave as they do, in health and disease, and how to bring the fruits of such a theory to the sickbed. To pursue the analogy further, we need a theory that stands to the germ theory as Einstein's theory of relativity stands to Newton's law of gravitation; or, to change the metaphor, as a simple steam engine stands to an atomic power plant.

That theory, of course, has yet to be announced. There are reasons why I think it *will* be—perhaps within the lifetimes of some of us.

Let us look at the difficulties that stand in the way, for they are formidable. Some of them we have already touched on in earlier chapters.

Recall, for example, that we saw that surgeons in Florida found that hemorrhages in throat operations are far higher during the second quarter of the moon than at any other time. But no new theory came from that, so far as I have been able to discover. There was only the practical (and useful) conclusion that operations should be avoided during the second quarter of the moon.

Svante Arrhenius found periodic variations in the onset of bronchitis and epilepsy, and concluded that these variations were caused by fluctuations of the atmosphere's electrical charges. But nothing further came of that either, until Frank Brown, Jr., used it as background for his own theory on the influence of subtle pervasive environmental forces on living clocks. And as we have seen, most biologists in this field disagree with Brown.

Janet Harker found that clocks set out of phase with each other, when surgically implanted in cockroaches, led to cancer and death of the cockroaches. Franz Halberg and his research associates have for many years taken the working position that rhythmic phenomena are associated with cancer, and it would seem that a blending of these two lines of work should lead to seminal ideas. But so far, these have not been forthcoming. Actually, in answer to a question that I asked him not long ago, Halberg countered, "How can you be *sure* that an *insect* has cancer?"[2]

Colin Pittendrigh flashed a warning to pharmacologists to use great caution in timing the administration of drugs. A common empirical observation about insulin reactions appears to support this. Diabetic patients find that they are very sensitive to insulin at night, when as few as twelve units may give unwanted reactions, whereas many times that dosage may be tolerated during the day. However, the low tolerance at night may be partly due to the usually lower intake of food at that time. Otherwise, how have variations in our sensitivity to drugs affected our daily lives? Who has ever seen a warning label on a prescription bottle that read, "*Not* to be taken between 10 p.m. and 8 a.m.?"[3]

[2] Following a seminar lecture that Halberg gave at the University of California, Berkeley, on November 3, 1964: "A Spectrum of Low-Frequency Rhythms in Physiologic Function."

Actually, more attention has been given to the hypothesis that cancer is caused by a virus. Two Nobel laureates have been studying cancer from this working position: Wendell M. Stanley, as director of the Virus Laboratory at the University of California, Berkeley; and James D. Watson, as director of the Cold Spring Harbor Laboratory of Quantitative Biology, on Long Island, New York. It would be fascinating to know whether normal cells are more susceptible to virus attack during certain phases of the circadian cycle than during others, and if so, where research in that direction might lead.

[3] Labels warn us against taking amphetamine at night, but that is because amphetamine interferes with sleep; periodic sensitivity has nothing to do with it. Drugs that must be taken with meals have bad side effects than can be offset by food, but again, no periodic sensitivity effect is involved.

Currently, more purposeful studies are in progress on the periodic susceptibilities of *insects* to chemical substances. William N. Sullivan, of the Entomology Research Division of the U.S. Department of Agriculture

Roland Fischer, of the Division of Behavioral and Neuro-biological Sciences, College of Medicine, Ohio State University, is studying the hallucinations that are induced by drugs. He finds that hallucinogenic drugs have a profound effect on a subject's perception of time and space, and this has led him to a critical examination of what time and space "really are." He holds that they are sharply different from what we ordinarily conceive them to be, and for his own operational purposes, he defines space as "data content" and time as "rate of data processing." These concepts are on the far edge of a challenging new approach to the problems of mental illness.

Finally, the list of diseases that either are, or that may become, cyclical seems endless. Clinical data on more than a thousand patients suffering from various periodic physical and mental illnesses were gathered by Curt Paul Richter, professor emeritus of psychobiology at the Johns Hopkins Medical School, during the many decades of research that he began in 1919.[4]

Among the organs that he found to be affected by these diseases were the joints, the bone marrow, the lymph glands, the stomach and duodenum, the peritoneum, the salivary glands, the sweat glands, the kidneys, the eyes, the skin, and the brain. The diseases that affected these organs ranged from simple swelling and pain to peptic ulcers, migraine, and epilepsy.

The mental and emotional states that he found to occur in periodic illnesses included elation, excitement, mania, depression, paranoia, and stupor; and these often went hand-in-hand with insomnia or excessive sleep, abnormal craving for food, periodic irresistible craving for alcohol, imaginary ill health, hallucinations, and even actual changes of personality.

To compound the problem, Richter observed that most

---

at Beltsville, Maryland, has found that flies are two or three times more susceptible to insecticides in the late afternoon than they are in the morning. This lead promises much-improved control of such pests as boll weevils, mites, and roaches.

[4] Curt Paul Richter: *Biological Clocks in Medicine and Psychiatry* (Springfield, Illinois: Charles C. Thomas, Publisher; 1965), p. 73.

illnesses do not show themselves to be periodic at the outset, but only after many months or years; and that they are thus ". . . readily overlooked by physicians."

Physicians today, fortunately, have ready access to the world literature on biorhythms through their own professional publications. What later was to become a most valuable communications medium for medical men began with a series of laboratory experiments on the liver function of rabbits.

In 1927 the young Swedish medical student Erik A. Forsgren recognized that the most important constituents of rabbit bile can be precipitated by barium chloride, and saw that he had at hand a method for measuring the amount of bile in a rabbit's liver at any time he might choose. In the laboratory he found that the amount of bile varied with time, and further, that it alternated with the amount of glycogen stored in the liver. "There is normally an antagonism in the liver cells between bile and glycogen formation," he reported. "When a liver cell contains an abundance of bile constituents, glycogen occurs sparingly, and vice versa. Thus liver function is periodic, bile formation alternating with the formation of glycogen."[5]

Within less than two years after Forsgren's finding, Jakob Möllerström, who was also working in Sweden, discovered that very large periodic variations occur in the sedimentation rate, a laboratory test that depends on the speed at which red blood cells settle to the bottom of a column of citrated blood, and that is widely used to diagnose the progress of both chronic and acute infectious diseases. In one case of cancer of the liver, Möllerström called the variations of the sedimentation rate "explosive" when he found more than 150-fold differences in the course of a single day. In summarizing his findings, he emphasized the danger of making any diagnosis from a single determination; and beyond this practical conclusion, he pointed

---

[5] Erik A. Forsgren: "On the Relationship Between the Formation of Bile and Glycogen in the Liver of Rabbits," *Skandinavisches Archiv für Physiologie*, Vol. LIII (1928), pp. 137–56.

Using a Control Data 3200 computer, medical researchers at the Biomedical Computer Simulation Laboratory at Salt Lake City, Utah, study mathematical models of circulatory systems, including heart rate, blood pressure, and related rapidly interacting physiological parameters.

to fundamental causes when he said, ". . . the sedimentation rate is of further interest when it becomes apparent that behind this phenomenon there are hidden biological reactions."[6]

These two findings started a ferment of excitement in Swedish medical circles; the next few years saw a sharp increase in research in related areas, along with a mounting interest in forming an international society devoted to studying biological rhythms from a medical point of view. The society

[6] Jakob Möllerström: "Om Kortperiodiska Förändringar i Blodets Suspensionsstabilitet," *Hygiea* (Stockholm), Vol. XCI (1929), pp. 497–508.

was formed in 1937, and by 1965 one of its founders, the German physician A. Jores, was able to write:

> In 1927 Forsgren discovered the periodic alternating storage of glycogen and bile in the liver, thereby opening up a wide field of 24-hour-periodic phenomena to investigation. Thus a new branch of science unfolded, biological rhythm research, which slowly acquired manifold interconnections with the other biological sciences, including medicine. In 1937 seven scientists—five physicians, one botanist, and one zoologist—founded an international society for the study of biological rhythms. Since then this new and interesting scientific discipline has spread all over the world. As is so typical of our time, the accumulated wealth of observations and experiments is well-nigh impossible for the single scientist to grasp.[7]

A very human note on the founders' first gathering at the Donneby health resort in Sweden was reported by Arne Sollberger, a Swedish physician who joined the Society later, and who for many years served as its secretary. "During happy and very rhythmic sessions," he wrote, "partly executed in the water, these scientists constituted the International Society for Biological Rhythm. Through Jores, the proceedings appeared in the *Deutsche medizinische Wochenschaft,* [and] later conferences of the Society were held at Utrecht, the Netherlands (1939), Hamburg (1949), Basel (1953), Stockholm (1955), Semmering, Austria (1957), Siena, Italy (1960), and Hamburg (1963)."[8]

One gigantic task remained. That was to sift and classify the "accumulated wealth of observations and experiments . . . well-nigh impossible for the single scientist to grasp," and make it accessible in a single reference volume. By the end of 1964 this had been accomplished by Arne Sollberger, the hard-working secretary of the Society for Biological Rhythm, who

---

[7] A. Jores, in "Forward" to Arne Sollberger: *Biological Rhythm Research* (Amsterdam: Elsevier Publishing Co.; 1965), p. vii.
[8] Arne Sollberger: *Biological Rhythm Research* (Amsterdam: Elsevier Publishing Co.; 1965), p. 308.

was at that time associated with the University of Puerto Rico Medical School.[9] Today the book serves researchers in many fields, for Sollberger extended his treatment of rhythm theory to include such esoteric specialties as general cybernetics, servomechanisms, energy theory of oscillators, basimetry and the law of initial values, physical and mathematical models, and mathematical-statistical analysis of time series.

Who, then, among tomorrow's biologists and medical men, will be so rigorously trained as to be able to put such sophisticated tools to dynamic use? How soon will they make the hoped-for breakthrough in the theory of living clocks?

Let us see.

During the last few centuries science and technology[1] have been advancing at a geometric rate, that is, at a rate proportional to some power of the length of time that has passed. Thus if progress in medical knowledge from A.D. 1950 to A.D. 2000 should turn out to be double that in the preceding half-century, the rate of advance would be called geometric. Every sign today points to progress at just such a rate, or faster.

What then of the men and women who will lead this drive?

In the half-century from 1800 to 1850 the United States boasted a mere 12,000 scientists and engineers. From 1850 to 1900 the number rose tenfold, to 102,000. From 1900 to 1950, *that* number increased tenfold again, to more than one million.[2] Whether or not our educational system can provide the extrapolated ten million technical men and women between 1950 and 2000 is an open question, but whatever the final score is, it will be very large.[3]

---

[9] Sollberger: *Biological Rhythm Research.*
[1] In the major systems for classifying knowledge, medicine is regarded as a branch of technology, that is, as one of the disciplines that apply the learnings of science to practical and useful ends.
[2] Harrison Brown, James Bonner, and John Weir: *The Next Hundred Years* (New York: The Viking Press; 1957), p. 173.
[3] The Council for Financial Aid to Education has brought this home

The quality of scientific and technical education is also rising apace. In preparation for medical research, for example, an entirely new concept is emerging. A number of major universities, both in the United States and abroad, are offering undergraduate curricula in bioengineering. The bachelor's degree in this field requires a rigorous series of courses in chemistry, physics, mathematics, computer science, biology, the humanities, and of course, engineering science. The holders of these bioengineering degrees will be ideally prepared to go on to graduate work, either toward the Ph.D. in engineering science, or toward the M.D.—in either case with careers in medical research in view.[4] The contributions that men and women so trained will make to the conquest of disease are incalculable.

Our future holds greater promise for the health of mankind, I firmly believe, than anything that we now dream of— a future, moreover, that lies very close at hand.

---

with their assertion that "80% of the scientists and engineers who have ever lived are alive today." *Saturday Review*, Vol. XIII (1970), p. 81.

[4] Charles Süsskind: "Bioengineering," in John R. Whinnery, ed.: *The World of Engineering* (New York: McGraw-Hill Book Company; 1965), pp. 114–29.

A. G. Pate: *Bioengineering: A Guide to Sources of Information in a New Interdisciplinary Activity* (Manchester, England: The Manchester College of Science and Technology; 1966), 12 pp.

Index

A Note About the Author

Ritchie R. Ward was born in Medicine Lodge, Kansas, in 1906. He received his B.S. in chemistry and a Master's degree in journalism from the University of California at Berkeley. After having spent many years as a chemist and laboratory manager in Hawaii and California, he retired in 1962 and began a new career as a writer and instructor and consultant on technical writing. He is the author of *Practical Technical Writing*, a textbook published in 1968, and has contributed biographical sketches of prominent scientists and engineers to *Current Biography*, published by H. W. Wilson. He also has prepared a seminar on technical writing for closed-circuit TV, originating at Stanford University and available to sixteen local industrial facilities. Mr. Ward has been a member of the U.S. National Committee for the International Federation for Documentation, a committee of the National Academy of Sciences, on which he served from 1961 to 1964. He is married, and lives in Orinda, California.

A Note on the Type

This book was set in Caledonia, a type face designed by W(illiam) A(ddison) Dwiggins for the Mergenthaler Linotype Company in 1939. Dwiggins chose to call his new type face Caledonia, the Roman name for Scotland, because it was inspired by the Scotch types cast about 1833 by Alexander Wilson & Son, Glasgow type founders. However, there is a calligraphic quality about Caledonia that is totally lacking in the Wilson types. Dwiggins referred to an even earlier type face for this "liveliness of action"—one cut around 1790 by William Martin for the printer William Bulmer. Caledonia has more weight than the Martin letters, and the bottom finishing strokes (serifs) of the letters are cut straight across, without brackets, to make sharp angles with the upright stems, thus giving a "modern face" appearance.

This book was composed and bound by The Haddon Craftsmen, Inc., Scranton, Pa., and printed by Halliday Lithograph Corporation, West Hanover, Mass.

Typography and binding design by The Etheredges.